Integrated Science

Michael E. Brint · David J. Marcey
Michael C. Shaw

Editors

Integrated Science

New Approaches to Education

A Virtual Roundtable Discussion

 Springer

Editors

Michael E. Brint
Uyeno-Tseng Professor
  of International Studies and
Professor of Political Science
California Lutheran University
Thousand Oaks, CA, USA
brint@clunet.edu

David J. Marcey
Fletcher Jones Professor
  of Developmental Biology
California Lutheran University
Thousand Oaks, CA, USA
marcey@clunet.edu

Michael C. Shaw
Professor of Physics and Bioengineering
California Lutheran University
Thousand Oaks, CA, USA
mcshaw@clunet.edu

ISBN: 978-0-387-84852-5   e-ISBN: 978-0-387-84853-2
DOI 10.1007/978-0-387-84853-2

Library of Congress Control Number: 2008937210

Printed on acid-free paper

springer.com

Preface

"Every now and then I receive visits from earnest men and women armed with questionnaires and tape recorders who want to find out what made the Laboratory of Molecular Biology in Cambridge...so remarkably creative. They...seek their Holy Grail in interdisciplinary organization. I feel tempted to draw their attention to 15th century Florence with a population of less than 50,000, from which emerged Leonardo da Vinci, Michelangelo, Raphael, Ghiberti, Brunelleschi, Alberti, and other great artists. Had my questioners investigated whether the rulers of Florence had created an interdisciplinary organization of painters, sculptors, architects, and poets to bring to life this flowering of great art? Or had they found out how the 19th century municipality of Paris had planned Impressionism, so as to produce Renoir, Cezanne, Monet, Manet, Toulouse-Lautrec, and Seurat? My questions are not as absurd as they seem, because creativity in science, as in the arts, cannot be organized. It arises spontaneously from individual talent. Well-run laboratories can foster it, but hierarchical organization, inflexible bureaucratic rules and mountains of futile paperwork can kill it. Discoveries cannot be planned; they pop up, like Puck, in unexpected corners."

— Max Perutz, in *I Wish I'd Made You Angrier Earlier* (1998)

The seminal discovery of Max Perutz, a method for phasing the X-ray diffractions from a protein crystal, provided the means for the calculation of atomic structures of macromolecules. This remains one of the most stunning achievements of interdisciplinary science. It is noteworthy that Perutz's early work, which transformed modern Biology, was carried out at the Cavendish Laboratory, a Physics laboratory in Cambridge that also yielded the remarkable interdisciplinary collaboration of Perutz's doctoral students, James Watson and Francis Crick.

Although the editors of this volume agree wholeheartedly with Perutz's view that the ultimate sources of scientific advances are found in individual perspicacity, we also recognize that institutional features that foster genuine integration of traditional scientific disciplines, like those existing in Cambridge at the Cavendish and later at the Laboratory of Molecular Biology, are essential to meet the needs of 21st century science. Indeed, the emergence of new scientific fields at the intersections of traditional, scientific disciplines and the increasing dependence on multidisciplinary approaches to solving problems at the frontiers of science demand responses and reformations at the institutional level.

Our goal in creating this "virtual roundtable" of discussants on the topic of integrative science, most decidedly, is not to attempt to provide a "Holy Grail in interdisciplinary organization." Recognizing that reform efforts are likely to be as varied as the institutions in which they occur, we have attempted to assemble, in a rather novel format, a symphony of voices that address the pluralistic nature of approaches to institutionalizing integrative science.

A few words about the virtual roundtable format of this book. As an enterprise, its goal is to synchronize the asynchronous: to assemble eminent thinkers on the subject of integrative science. The "participants" come from different perspectives and experiences, and include Nobel Laureates, University Presidents, serious scholars, and distinguished scientists. Although their comments, talks, articles, and interviews on this subject may have taken place at different times and in widely different venues, we have collected them into an organized, coherent ensemble of integrated conversations about the necessity, promises, challenges, and implementation of integrative approaches to scientific research and education. We have chosen to frame the roundtable conversations by posing a series of central questions. We hope that the answers to these questions will be of interest to a wide range of scientists, educators, and university and college administrators facing the exciting, if daunting, hurdles involved in integrative reform. The discussions of the questions are certainly not meant to be comprehensive. Rather, we asked 10 of the most pressing questions related to integrative science and sought answers from 21 of the world's experts on the subject. At times, their voices are mutually reinforcing. In other instances, divergent answers to the same question arise, a sign of the timeliness and vigor of discussions on integrated science.

The book is divided into three parts. The first and second parts focus on integration at a large, structural level. Here, integration refers to the relationship between academic science and government (Part I) and between academic science and industry (Part II). Throughout these discussions, a second form of integration emerges. Academic science itself is seen as increasingly interdisciplinary – depicting a convergence of disciplines often resulting in new fields of study – or multidisciplinary – an approach that emphasizes the integration of disciplines employed to solve specific problems. The final part of this work analyzes the implications of interdisciplinary and multidisciplinary approaches to modern scientific investigation and education.

Former Vice President Al Gore begins the discussion with the intriguing notion of "distributed knowledge" – a metaphor drawn from computer science. Of critical importance to this distributed system, he emphasizes scientific literacy among policy makers and politicians. If we are seriously to confront global issues such as climate change, Mr. Gore argues, we must have policy makers who are part of the distributed knowledge system of science that emanates from, in large part, the universities and cycles through the government. On a global or macro-level, the promise of integrated science is accompanied by a grave sense of urgency, according to both Mr. Gore and Dr. Bruce Alberts, former president of the National Academy of Sciences. "Today," Dr. Alberts says, "we find it difficult to meet the basic needs of the Earth's six billion people. How, then," he asks, "can we hope to meet the basic needs of the nine billion people expected to inhabit our planet by 2050?"

Dr. Elias Zerhouni, Director of the National Institutes of Health, also encourages a new, more integrative, structure of organizing science to respond to the discovery of a unifying set of "principles that link apparently disparate diseases through common biological pathways and therapeutic approaches." In his discussion, he guides us through the NIH Roadmap effort that includes the support of novel, interdisciplinary, organizations of research teams and grants awarded to high risk scientific enterprises.

Dr. James Duderstadt, former President of the University of Michigan system, analyzes the convergence of government, academic science, and private industry. Specifically, he provides an overview of the ramifications of the pivotal Bayh-Dole Act of 1980, which engendered a fundamental shift in the ways in which technology transfer of academic research occurred. Whether the diverse fields of integrative science in the academy lend themselves more or less to the guiding hands of industry remains to be seen. Dr. Duderstadt warns that the traditional values of the academy must be preserved while institutions of higher education respond to the demands of the market place.

The discussion of integration in relation to industry and capital continues in the contribution by Dr. Stanley Aronowitz, Distinguished Professor of Sociology and Director of the Cultural Studies Program at the Graduate Center, City University of New York. Dr. Aronowitz argues for the de-comodification of the University. In contrast to Dr. Duderstadt's desire to maintain the integrity of traditional values of the university while responding to market realities, Dr. Aronowitz argues that the line is too often blurred between the idealized curriculum of the academy and the focused priorities of industry. Dr. David Kirp, Professor of Public Policy at the Goldman School of Public Policy is equally skeptical of such integration of education and industry. "While the public has been napping, the American university has been busily reinventing itself," Professor Kirp begins. The new shape of the university has been tailored to the demands of the marketplace.

Hank Riggs, Founding President of the Keck Graduate Institute, reflects on the roles of leadership in industry and higher education respectively. Having had experience in both, President Riggs suggests that although we should be mindful of their differences, leadership in these areas is surprisingly similar, particularly in respect to the challenges that both educational and industrial leaders confront. Dr. William Haseltine discusses the trends in science from a very different vantage point. As a former professor at Harvard Medical School and now entrepreneur, Dr. Haseltine likens disciplines within science as a wonderful tool set. But, he warns, innovation, discovery, and development demand that scientists have access to more than one single tool – one single disciplinary approach to solve problems.

Exploring the material and sociological factors involved in such interdisciplinary training, Dr. Steven Brint, Associate Dean and Professor of Sociology at the University of California, Riverside, offers a balanced account of whether integrative science is a passing fancy of the academy. From a reservoir of data, Dr. Brint reports on factors from technological change and federal and private funding projections to demographic trends and global competition that may determine whether new directions in science will have a lasting impact on the landscape of higher education. Dr. Paul Grobstein addresses some of the challenges that academics confront in developing these new directions in science education. In terms that reflect evolutionary psychology, he likens disciplines to tribes who express an inclination to share observations and stories only with people who are in some sense "like themselves."

The implication of the integration of science at the undergraduate level is tackled first by Dr. William Wulf, former President of the National Academy of Engineering, in his discussion of question six. Namely, that a major change is occurring, albeit gradual, beginning with a re-definition of "the fundamentals" through to an articulation of faculty motivations and incentives for gaining practical experience in industry. Dr. Donald Kennedy, President Emeritus at Stanford University, continues by succinctly describing the competition between depth and breadth in undergraduate science education. He also enumerates the inexorable fiscal challenges associated with the capital-intensive nature of science education at the undergraduate level. He concludes with a concise and insightful summary of the obstacles which must be overcome in supporting undergraduate faculty. Together, these two essays capture the essential benefits, opportunities and difficulties in world-class, undergraduate integrative science education.

Dr. Kennedy and Dr. Rita Colwell, former director of the National Science Foundation, then discuss whether new directions in scientific training encourage a more diverse body of scientists. Both point to recasting science training as fundamental to the flourishing of diversity. "The interconnectedness of life is a very deep law," Dr. Colwell remarks, "and greater diversity makes for a more robust ecosystem than does a monoculture. The environment must nourish any organism, or it will not survive – just like the environment for a young scientist, which can be chilling or nurturing." Dr. Kennedy suggests that just such an environment can be found in liberal arts institutions where one-on-one mentoring is part of the institutional culture.

In question eight, Dr. Colwell takes up the issue of graduate training. She provides a broad view of new directions at the Master's level with focus on professional training rather than preparation for the Ph.D. As an example of such a program, Dr. Colwell refers to the Professional Science Master's (PSM) degree, a program supported by the Alfred P. Sloan Foundation. Until 2005, the National Outreach Coordinator for the Sloan Science Master's Initiative was Ms. Sheila Tobias. Ms. Tobias discusses the details of this new approach; an approach that integrates elements of industry and education, emphasizes interdisciplinarity, and subsequently changes the goals of the traditional Master's Degree for those students seeking work in the science industry.

In response to a question about the challenges of training interdisciplinary Ph.D.s, Dr. David Baltimore, a Nobel Laureate, and past president of both Rockefeller University and The California Institute of Technology, provides a historical perspective that highlights the important role of combining technology and instrumentation with molecular biology. He then sketches out the implications of this paradigm for the future training of practitioners of integrative science, and suggests institutional changes that will enhance this training. In responses to the same question, Drs. Golde and Gallagher of the Carnegie Foundation for the Advancement of Teaching provide a cogent discussion of obstacles that doctoral students face if they wish to conduct interdisciplinary research. Finally, Drs. Cech and Rubin, of the Howard Hughes Medical Institute (HHMI), describe the considerations that led to the *de novo* establishment of an interdisciplinary research institute, the Janelia Farm campus of HHMI.

The volume ends with Robert Venturi offering a short course on the philosophy and grammar of space. He applies these concepts to the design of science buildings. In this discussion, he articulates a new vocabulary for creating scientific space for the 21st century. Dr. Claire Fraser offers her observations on developing the building plan for the Institute for Genomic Research in Rockville, Maryland. In conclusion, she claims, more by luck than design, the proximity of scientists created the relationships needed for the integrated science that the Institute sought to establish.

We hope that the reader finds the roundtable discussions stimulating, and that some reformative utility will be found in the viewpoints contained herein. The roundtable is intended as a launching pad for further discussions amongst colleagues who are focused on promoting integrative approaches at a variety of institutions. If this volume stimulates even a modicum of such, we will be satisfied with our efforts.

California, USA   Michael E. Brint
                                                                             David J. Marcey
                                                                             Michael C. Shaw

Contents

Part I
The Promises and Challenges
of Integrated Science

Chapter 1
In What Ways Can or Should Science and Government Be Integrated?

Al Gore, Former Vice President, United States of America (1993–2001); Nobel Laureate

Albert Gore, Jr. was the 45th Vice-President of the United States (1993–2001). Gore previously served in the U.S. House of Representatives (1977–85) and the U.S. Senate (1985–93). A prominent environmental activist, he shared the 2007 Nobel Peace Prize with the Intergovernmental Panel on Climate Change.

M.E. Brint et al. (eds.), *Integrated Science*, DOI 10.1007/978-0-387-84853-2_1,
© Springer Science+Business Media, LLC 2009

You have spoken of changing the fundamental metaphor for integrating science and government to address some of the salient issues of our time including global warming. What metaphor would you use to describe this integration?

Not too long ago, the metaphors of science migrated easily to the realm of political and economic affairs. In previous generations, the logic and lingo of science – from Newtonian physics to the industrial science of Frederick Taylor – informed our public conversation. But not today – or at least not very often. When I say that our current, chaotic political culture reminds me of Ilya Prigogine – that because our system has more and more energy coming in, it will eventually reorganize itself into a complicated and unpredictable new system . . . nobody has a clue what I'm talking about.

As a result, the language we use to discuss public problems is less vivid and less robust than it ought to be. Chaos theory may offer clues for when government should intervene in the economy. Economic policy perhaps should focus less on "priming the pump" – and more on "imprinting the DNA." Evolution could offer insight into our social structures. But at the moment, we lack the vocabulary to even begin such discussions. We either avoid scientific metaphors altogether – or we lean against the crutch of Industrial Age metaphors that are splintering with age. In particular, we continue to rely on the metaphor of the factory – of mechanized mass production – well after it has exhausted much of its supportive force.

[I]n the spirit of academic inquiry, let me propose an alternative metaphor . . . an updated metaphor . . . a metaphor more appropriate to the times and more muscular in its power to explain. It is the metaphor of distributed intelligence.

In the beginning of the mainframe computer era, computers relied almost totally on huge central processing units surrounded by large fields of memory. The design was much like a mass-production factory. The CPU would send out to the field of memory for raw information that needed to be processed, bring it back to the center, do the work, and then distribute the answer back into the field of memory. This technique performed certain tasks well – especially those that benefited from a rigid hierarchy or that depended on the outer reaches only for rote tasks.

Then along came a new architecture called massive parallelism. This broke up the processing power into lots of tiny processors that were then distributed throughout the field of memory. When a problem was presented, all of the processors would begin working simultaneously, each performing its small part of the task, and sending its portion of the answer to be collated with the rest of the work that was going on. It turns out that for most problems, this approach is more effective.

But somehow this idea, revolutionary as it was in the computer world, never traveled to other regions of our life – and didn't come anywhere near politics. And that's a shame. Because in the realm of politics or economics or public policy, the metaphor of distributed intelligence has enormous explanatory power. It offers an insight into why democracy has triumphed over governments that depended exclusively on a central authority. And it illuminates why private sector organizations are shedding their middle layers and pushing power, information, and influence to frontline workers. Taken a step further, it even helps explain phenomena as diverse as virtual communities on the Internet and television programs like "America's Funniest Home Videos."

How does distributed intelligence relate to academic science and interdisciplinary approaches to science?

...At their best, the scientific community and the university community embody the ideal of distributed intelligence. The great power of science derives in part from specialization into disciplines. But much of the power also comes from open criticism and communication across disciplines. Indeed, some of the most significant discoveries have emerged from the productive friction that occurs when different perspectives rub against each other and produce the spark of new insight. But if the physicists don't talk to the chemists, and the chemists don't talk to the economists, and if the economists don't talk to the climatologists, then distributed intelligence is more aspiration than reality.

What role should distributed intelligence play in shaping public policy? Why hasn't the idea caught on in Washington?

[T]he notion of distributed intelligence has not migrated to our public conversation [because of] the growing disconnect between science and democracy. Walk through the halls of Congress, and you'll see the Gucci loafers of corporate lobbyists, but not the white lab coats of American scientists. Page through a directory of members of Congress, and you'll find well over 150 lawyers, but only 6 scientists, 2 engineers, and 1 science teacher among the 535 people in the House and the Senate.

As a result, scientific concepts sometimes elude the vast majority of our elected officials. That is inherently unfortunate, because we want well-rounded leaders.

But let me dwell a moment about some of the harder-edged consequences in the hopes that it will solidify my case for this new metaphor: Lack of scientific understanding undercuts support for the pursuit of further understanding, which fosters deeper ignorance, which in turn further erodes support for battling that ignorance. It's a vicious cycle . . .

For much of this century, Americans have benefited from a virtuous circle – a virtuous circle of science and success. As the nation generated wealth, a portion of that wealth was invested in research, science, and technology. Those investments helped answer what seemed answerable – and eventually spawned still greater wealth, which was then invested in still more research. On and on it went. In this virtuous circle – launched with bipartisan agreement – prosperity generated investment, investment generated answers, and answers generated further prosperity.

But now – because of the woeful lack of knowledge . . . that virtuous circle risks coming undone. At the very moment a new age demands continued investments in science and technology, there are some in Congress threatening to turn the clock backward with the largest cuts in 15 years. . . . At the very moment global economic competition and global environmental degradation demand civilian research and the technologies it often produces, this Congress is proposing the sharpest cuts in non-defense research since America was fighting World War II....Research on issues that will affect the health of our children, the condition of our planet, and the vibrancy of our economy risks being slashed to the bone.[1]

- Global warming...down.
- Supercomputers...down.
- Nuclear non-proliferation...down.
- New materials...way down.
- Solar energy...way down.
- Environmental satellites...down.
- Water quality...down.

It's like they're living in a gravity-defying universe. Everything that ought to be up is down. Everything that ought to be open is closed. Their science policy is straight out of science fiction. A few may talk like Johnny Mnemonic, but most support policies designed for Fred Flintstone...And right now, several agencies – in particular, the National Science Foundation – are sputtering along with stopgap funding that makes it almost impossible to plan and difficult even to finance day-to-day activities.

[At this time,] we have a choice of two paths. One path retreats from under-standing, flinches in the face of challenges, and disdains learning. It leads to a know-nothing society in which the storehouses of knowledge dwindle, the spigots of discovery are turned off, and missions of exploration are stalled on the ground. This society bases regulations on suspicion instead of science, says that DDT isn't harmful, and claims that global warming is the empirical equivalent of the Easter Bunny. But there's another path – infinitely brighter and considerably more Ameri-can. It leads to a learning society whose government continues to fund basic science and applied technology and in which the virtuous circle of progress and prosperity is alive and functioning. And it's a trail that's within our power to blaze.

We have in our hands and minds and souls the power to create this learning society, which harnesses the power of distributed intelligence and uses it to improve our lives.

---

[1] Editor's Note: According to the American Association for the Advancement of Science (AAAS), "Federal research investments are shrinking as a share of the U.S. economy, just as other nations are increasing their investments. [T]he federal R&D investment exceeded 1 percent of U.S. Gross Domestic Product (GDP) until recently, buoyed by big increases in weapons development, but is now declining sharply. Federal investments in development, mostly in DOD, have held steady as a share of the economy, but the federal research/GDP ratio is in free fall down to a projected 0.38 percent in 2008, below the long-term historical average of 0.4 percent after gains in the late 1990s." "Federal Research Would Continue to Fall in 2008 Budget," February 2007, AAAS, http://www.aaas.org/spp/rd/guihist.htm.

Bruce Alberts, Former President (1993–2005), U.S. National Academy of Sciences

Dr. Alberts is a respected biochemist recognized for his work both in biochemistry and molecular biology. He has spent his career making significant contributions to the field of life sciences, serving in different capacities on a number of prestigious advisory and editorial boards, including as chair of the Commission on Life Sciences, National Research Council.

Vice-President Gore argued that there are two paths that we can follow: One leads to greater integration of science and government, the other to increased separation that will reduce our potential for dealing with the global issues that we are confronting. Dr. Alberts, how can the organization of science be most effective in terms of its social impact?

Science and scientists must attain higher degrees of influence both within their nations and throughout the world – not just for the sake of their professional well-being but for the well-being of the societies in which they live and work. To accomplish this goal, we must take full advantage of all the tools that we have at our disposal – most notably, the new information and communication technologies that offer tremendous opportunities for us to be more effective as scientists, scientific advisors, and science communicators.

What are some of the challenges to attaining this goal?

The barriers we face in achieving this goal are many. Leadership, of course, is critical not just among researchers in our own disciplines but among science administrators and public officials in the larger political community. We must develop an image of science, moreover, that the public finds welcoming, not frightening. The most effective way to alter public perceptions over the long term is through science education for children based on a curriculum that provides hands-on learning experience; rewards team work; and encourages teachers to coach, not dictate. A new framework for science education can only be constructed if the best scientists – and the most prestigious scientific institutions – get behind the effort and clearly indicate that providing every student with a basic understanding of science may be more important for society as a whole than for science itself. This effort will likely require a change in attitude among scientists who for too long have erected barriers between their expert communities and the larger world in which they live and work.

What is the driving force behind your advocacy for global science?

[...] Today we find it difficult to meet the basic needs of the Earth's six billion people. How, then, can we hope to meet the basic needs of the nine billion people expected to inhabit our planet by 2050 – a population that will include not people living in another time, but many of our own children and grandchildren?

What role does multidisciplinary or interdisciplinary science play in answering this question?

The answer lies not simply in science but in a particular science that is place-based, multidisciplinary, and acknowledges the importance of social sciences. In other words, a science that defines both purpose and excellence in a broader context than has previously been the case.

You have advocated for "global science." What efficacy has this movement demonstrated and what challenges still remain?

Global science has made great advances over the past several decades. But, as recent controversies over the underlying forces driving the HIV/AIDS epidemic in South Africa and the value and risks of distributing genetically modified food in Zambia show, global science can have only so much influence over national and local decision-making. The fact is that you cannot rely on outside scientific advice to sway policies within a nation. Each country must develop its own indigenous scientific expertise and capacities, not only to increase the likelihood that policy makers will pay closer attention to scientific findings, but also to create an effective infrastructure for adapting global scientific knowledge to national and local needs.

An important aspect of the overall mission of science is to integrate scientific findings into policy. Politicians concentrate on short-term goals not because they are short-sighted, but because the environment in which they operate demands such a focus. In contrast, science, by concentrating on long-term impacts and goals, can provide a countervailing perspective for policy makers that can help them gauge the lasting effect of their decisions on their citizenry. Individual scientists rarely can achieve a high level of influence either within their disciplines or their larger societies solely on their own. They must have the support of effective, powerful institutions. Development of such institutions has been the hallmark of science in the North and it must become one of the principal strategic elements of science-based development in the South as well. All those involved in the promotion of science in the developing world – including national governments, international organizations, donors, and global scientific institutions – must focus on institutional capacity building as a major aspect of their efforts.

In regard to integrating science, then, you are suggesting that solving social problems requires bringing together an array of disparate social, political, and scientific persons and institutions. What role in particular do educational institutions play?

Universities play a central role in the training of the next generation of scientists. Research centers, often in partnership with universities, help to tie scientific research to critical societal issues and thus assist in broadening the reach of scientific endeavors. Scientific academies – self-governing institutions managed directly by their membership – enjoy a set of distinct advantages based on their stability, merit-based principles, and political independence. Such enviable characteristics should provide academies with the opportunity to enjoy a powerful voice in science-based policy discussions within their own nations.

With your experience as President of the National Academy of Sciences, what role do you see such academies playing in global science?

Scientific academies . . .have been largely underutilized within their societies. It is in the interest of both scientific communities and the societies in which they function to seek ways to enable science academies to develop a larger role – including becoming respected advisors to their own governments on science-related concerns. That has been the goal of the InterAcademy Panel for International Issues (IAP). The panel, whose roots lie in an unprecedented meeting of the world's science academies in New Delhi in 1993, is dedicated to building the capacity of academies in ways that enable them to become more powerful voices within their own societies. Today, 88 academies – virtually every merit-based national science academy in the world – is a member of IAP, whose secretariat is housed under the administrative umbrella of the Third World Academy of Sciences (TWAS). IAP's efforts have been based on the simple assumption that, by sharing information and experiences and engaging in programs designed to build skills and expertise, each institution can grow stronger in its own efforts to gain greater visibility and influence within its society.

What role do you see the US National Academy of Sciences (NAS) playing in these efforts?

The US National Academy of Sciences (NAS) has been an enthusiastic supporter of IAP's efforts. Small and impoverished nations, in particular, cannot be expected to have either the expertise or the financial resources necessary to conduct comprehensive scientific research to address all of the critical issues that they face. Providing free access to scientific data and information generated by scientists in other countries could help scientists working under difficult circumstances to overcome some of the constraints that they face. It is for this reason that the US National Academies have made the more than 3000 scientific reports and books at www.nationalacademies.org freely available as pdfs to anyone in over 140 nations.

Science has always been an international enterprise. But by pooling information, in the age of rapid information exchange, science could offer the added benefit of helping all nations – rich and poor, big and small – make wise decisions based on the current state of expert knowledge.

What role do you see the United States scientific community playing in these efforts?

Most US scientists do not know anything about Brazil, China, or India, let alone Cambodia, Chile, or Senegal – and they often know next to nothing about the scientific communities in these nations. It's not that our scientists are disinterested in these places and the science that is done there; it's simply that the education we receive and the jobs we hold don't usually focus on issues beyond our borders. If [we] can help lower the barriers of communication and greatly increase both understanding and the involvement of the US scientific community with scientific communities elsewhere, we will have made a significant contribution to both our professions and our societies. In fact, in the post 11 September world, spreading science and the values of openness and honesty that it embodies should be widely recognized as a noble endeavor that carries benefits well beyond the material well-being that results from science-based economic development.

Dr. Alberts, Vice-President Gore spoke of the need to increase governmental spending on social issues. What changes would you like to see in the granting process?

The granting system is basically too conservative. One indication of the conservative nature of this system is that the age at which young scientists are getting funding has been gradually and continuously going up. In my generation, the assistant professors were hired and had their own research laboratories by the age of 27 or 28. Now it's more like 38. It really has changed dramatically in a short time. This means, first of all, that by the age you are funded, you have lots of responsibilities and probably, a family; you are not as energetic as you were before; but most importantly, you have had a long, long apprenticeship. Combined with the fact that if you want to continue getting funded then you better work on what you were working on before; this causes science to drill lots of deep holes in a limited number of areas.

Every time I go back to writing my textbook, I get to think broadly about biology (cell biology) and I realize how many areas are not being investigated; many of them at the interface between medicine and cell biology. I also realize how many people are digging in the same hole, competing with each other to beat the other laboratory by two weeks or a month. This is not the way I want to do science and I don't think it is the best way for the public to invest their money. So, we badly need a system that is much more encouraging of new ideas. I have been on numerous committees, one of them for the NIH, which put out a report suggesting that for first-time investigators, no preliminary results be allowed. There should also be special review groups for these people who are just setting up their lab for the first time. We want a system in which people with new ideas are rewarded. Of course this demands a lot of choice. You don't want to give every young person an NIH grant or NSF grant. But the best young people should really have a chance to do something different.

I think we need a different mindset concerning what we are looking for in our granting system. And, of course, interdisciplinary research is an important part [of that new mindset.] It often gets undervalued because it does not fall into a slot – and it is harder to evaluate. There is not a community that understands these areas deeply enough. We have to do a lot better and pour more energy into rewarding innovation. Basically where the money is available drives what scientists do. Right now, young scientists get a message that if you are not applying for work that you have basically completed, forget about getting a grant. Clearly major changes are needed.

Chapter 2
What Are the Promises and Challenges of Scientific Integration?

Elias A. Zerhouni, M.D., Director, National Institutes of Health

Dr. Zerhouni leads the nation's medical research agency and oversees the NIH's 27 Institutes and Centers with more than 18,000 employees and a fiscal year 2006 budget of $28.6 billion.

M.E. Brint et al. (eds.), *Integrated Science*, DOI 10.1007/978-0-387-84853-2_2,
© Springer Science+Business Media, LLC 2009

Could you address the challenges and opportunities you see in terms of Integrated Science and the National Institutes of Health?

When future historians look back upon our era, they will see it as a pivotal moment, with humanity at a crossroads. Poised between nature, science, and a multiplicity of societies, we – the actors of tomorrow's historical accounts – look out upon an exciting but uncertain horizon. The completion of the Human Genome in 2003 has placed within our grasp a map of our own architecture – the "book of life." Computer technologies allow us to process unimaginably large amounts of data and share the results with each other, instantaneously, almost anywhere on the globe. In these and so many other instances, we live in a world "made flat" by our knowledge. [1] Yet, we need not look far to see that our successes have given rise to new challenges. Technologies that allow modern civilization to thrive are also warming the planet. World populations, too, are on the rise – particularly in places that economic "flattening" has neglected. All around the world, chronic diseases are becoming epidemic, and infectious diseases stubbornly resist our best efforts at containment. Our knowledge has, in the prophetic words of Francis Bacon, given us power: for better, and for worse.

Today, poised at a historic crossroads, we are finally in a position to actualize the second, if lesser known, part of Bacon's famous phrase: "Nature to be commanded must be obeyed." To preserve Nature – be it the environment, or human nature – we must follow Nature's rules. We may still have some way to go in our efforts to understand those rules in general, and the rules of living systems in particular. Yet, we appreciate more fully than ever before the complexity of those systems. We are beginning to recognize that we cannot simply follow the traditional paths of science – as successful as they have been – if we hope to understand nature and effectively address the many challenges we face. We must chart a new course, forge new paths. An integrated approach to science must be our guide.

In the effort to forge these new, integrated paths for biomedicine, the National Institutes of Health is uniquely situated to take a pioneering role. NIH is the world's pre-eminent medical research center and the steward of medical and behavioral research for the Nation. With a $29 billion dollar Federal budget, NIH supports peer-reviewed research at universities, medical schools, hospitals, and research institutions throughout the U.S. and the world. It conducts research in its own laboratories, trains research investigators, and disseminates science-based health information. Only a single institute in 1930, NIH evolved, with Congressional directives, to the 27 Institutes and Centers of the present day. NIH breakthroughs have transformed medicine, and NIH trainees or grantees have garnered 122 Nobel prizes.

NIH's achievements have been grounded in its successful adaptation of modern biomedicine's finest principles and practices. Each of its 27 Institutes and Centers focuses on a set of diseases, disorders, or bodily systems that reflect society's most

---

[1] Thomas L. Friedman, *The World is Flat: A Brief History of the Twenty-First Century* (New York: Farrar, Straus and Giroux, 2005).

pressing medical concerns. Within this "intramural" program, NIH researchers conduct cutting-edge, often long-term, research. Many draw upon the NIH Clinical Center – the Nation's largest hospital devoted entirely to clinical research – in their efforts to bring the latest laboratory discoveries to bear on a variety of intractable patient conditions. Moreover, the NIH extramural grants program, now in its 62nd year, relies on its unparalleled peer review system to determine the most promising research projects outside NIH, and to fund those projects with few constraints. The NIH peer review system, which continues to be a model the world over, has periodically been retooled to better fit changing times. Yet, at its heart, it continues to adhere to one of modern medicine's greatest lessons. In the words of Vannevar Bush – echoing Louis Pasteur: "The history of medical science teaches clearly the supreme importance of affording the prepared mind complete freedom for the exercise of initiative." [2]

Clearly, this bedrock of biomedical progress must be preserved. At the same time, today, efforts to prevent, detect, and treat disease demand that we better understand the dynamic complexity of the many biological systems of the human body and their interactions with our environment at several scales – from atoms, molecules, cells, and organs to body and mind. A dizzying array of parts participates in intricate and interwoven pathways that, together, contribute to health. Research is just beginning to converge on unifying principles that link apparently disparate diseases through common biological pathways and therapeutic approaches. Today, and in the future, biomedical research must reflect this new reality. Advanced technologies, including sophisticated computational tools and burgeoning databases, likewise span diseases and disciplines. The scale and complexity of our multi-faceted biomedical research problems increasingly demand that scientists move beyond the borders of their own disciplines and apply new organizational models for team science.

The fit between this emerging biomedical reality and traditional biomedical research – the latter resting as it does on individual researchers, distinct diseases, independent disciplines and institutions, and a curative approach to disease that intervenes only *after* symptoms have become manifest in a patient – is far from perfect. How, then, can NIH integrate these related but still divergent perspectives? Upon becoming the NIH Director in 2002, I devoted myself to working with my colleagues to find solutions. These solutions have taken the form of the NIH Roadmap for Medical Research.

How was the Roadmap constructed?

The NIH Roadmap began with a series of conversations: with directors of NIH Institutes, Centers, and programs; with legislators and members of the American public, who had supported a 5-year doubling of the NIH budget (completed in the 2003 fiscal year [FY]); with patient advocacy groups; and with scientific leaders at

---

[2] Vannevar Bush, *Science, the Endless Frontier* (Washington, D.C.: U.S. Government Printing Office, 1945).

institutions around the country. These conversations underscored the need for NIH
to reexamine its portfolio, with a goal of identifying critical scientific gaps that it
might help fill. Among these gaps was the difficulty that NIH itself faced when
it sought to support cross-cutting research programs that fell outside the scope of
any one NIH Institute or Center. More conversations – and discussions – followed.
My colleagues at NIH and I consulted with hundreds of nationally recognized leaders in academia, industry, government, and the public. Where, we asked, should
biomedical research be headed? What were the roadblocks that obstructed its path?
In answer to these questions, we introduced the NIH Roadmap in the fall of 2003.[3]

The Roadmap was organized around three major themes: New Pathways to Discovery, Research Teams of the Future, and Reengineering the Clinical Research
Enterprise. The three themes are mutually reinforcing and, as they develop, become
increasingly convergent. Working groups, each led by Institute directors and with
input from the NIH Council of Public Representatives and the Advisory Committee to the Director, determined an array of potential initiatives related to these
themes. Their suggestions were further refined and, from them, we chose the initiatives that would represent the first round of Roadmap initiatives in FY2004. We
strongly believed that their successful implementation would allow NIH to more
effectively support innovative and high-risk research, incubate new ideas, and stimulate the development of transforming strategies that could benefit the entire scientific
community.

Initially, each NIH institute contributed 1% of its budget to fund the Roadmap. In
December, 2006, Congress, recognizing the initial success and future promise of the
Roadmap, responded to these integrative efforts by authorizing a "Common Fund"
within the Office of the Director to provide stable support for Roadmap programs.
President Bush signed the "NIH Reform Act of 2006" into law in the following
month.

Teams

One of the Nation's most pressing challenges is to generate and maintain the trained
and creative biomedical workforce necessary to tackle the converging and daunting research questions of this century. NIH is actively experimenting with building
research teams of the future through the Roadmap.

Interdisciplinary Research

To lower artificial organizational barriers and advance science, the Roadmap established a series of awards that makes it easier for scientists to conduct interdisciplinary research. These new awards support an array of initiatives. They promote,
for example, the training of scientists in interdisciplinary strategies, the creation of
specialized centers that help scientists forge new, more advanced disciplines from

---

[3] Elias Zerhouni, "The NIH roadmap," *Science* 302 (2003): 63–64; 72.

existing ones, and the planning of forward-looking conferences with the potential to catalyze collaboration among the life and physical sciences – important areas of research that historically have had limited interaction.

The largest component of the Roadmap's interdisciplinary research program is the Interdisciplinary Research Consortia. Launched in 2007, the Consortia employ interdisciplinary approaches to medical problems that have proved resistant to solutions from single-, or even multi-disciplinary approaches. (Whereas *multi*disciplinary research involves teams of scientists approaching a problem from their own discipline, *inter*disciplinary research *integrates* elements of a wide range of disciplines so that all of the scientists approach the problem in a new way.) [4] Each consortium adopts one general medical problem for particular attention. Presently, Consortia members are investigating problems that range from ways to preserve fertility in women with cancer and methods for deciphering the basis of neuropsychiatric disorders to strategies for devising a coordinated and systematic approach to regenerative medicine and obesity. In addition to addressing these particular medical problems, Consortia members are contributing to the development of a medical culture that is more open to interdisciplinary teamwork.

In addition to funded initiatives, the Interdisciplinary Research initiatives include non-funded projects that aim to change NIH policies and procedures. Chief among these is a reconsideration of how NIH should best recognize leadership of collaborative projects. We have seen that NIH has followed the traditions of modern medicine by singling out one "Principal Investigator" (PI) as the guiding mind behind each award. This policy, however, effectively acts as a roadblock to funding truly interdisciplinary team projects. In response, NIH has moved toward recognition of *multiple* PIs for any award. Moreover, interdisciplinary research tends to resist fair evaluation by review groups that have largely been cast along standard disciplinary lines. Consequently, we have analyzed review strategies that are better able to assess interdisciplinary research proposals. These Roadmap projects have also helped inform the larger, recent NIH-wide effort to reconsider how we conduct the scientific review of all NIH grant applications.

Director's Pioneer and New Innovator Awards

At the same time, traditional review groups have on occasion displayed conservative tendencies that discourage certain kinds of *individual* investigators: particularly, those who propose pioneering projects considered to be high-risk, and those who are new to the field and consequently lack detailed substantiating data and a proven track record. The work of both groups is essential to the future of biomedical research and must be preserved and encouraged. Therefore, the Roadmap offers two

---

[4] Committee on Facilitating Interdisciplinary Research and the Committee on Science, Engineering, and Public Policy, National Academy of Sciences, National Academy of Engineering, Institute of Medicine of the National Academies, *Facilitating Interdisciplinary Research* (Washington, D.C.: The National Academies Press, 2004).

programs designed to achieve this end: the NIH Director's Pioneer Awards and the New Innovator Awards.

Winners of the highly selective Pioneer Award are scientists of exceptional creativity who propose highly innovative approaches to major contemporary challenges in biomedical research. By bringing their unique perspectives and abilities to bear on key research questions, these visionary scientists may develop seminal theories or technologies that will propel fields forward and speed the translation of research into improved health. Since the program started in 2004, there have been 47 awardees; already, their work is producing impressive, and potentially transformative, results. One awardee is using a multi-disciplinary approach to understand the principles governing T cell-mediated autoimmunity. The research could lead to new ways to predict or preempt autoimmune diseases such as multiple sclerosis or type 1 diabetes. Another is employing antigenic cartography to map differences in seasonal influenza strains worldwide: knowledge that could significantly improve our ability to track the influenza virus and select proper strains for vaccine preparation. [5]

The Director's New Innovator Awards support exceptional new investigators who have not yet received an NIH R01 grant, but who take particularly innovative approaches to biomedical or behavioral research. One Innovator awardee is researching the role of the *in utero* environment on the development of childhood obesity. Using state-of-the-art biological analysis technology, another Innovator awardee is developing a method for personalized diagnosis of a form of brain cancer known as glioblastoma multiforme. If validated, this technology could guide therapy decisions to the agents that would be most effective for the individual patient.

The creative scientists we recognize with NIH Director's Pioneer Awards and NIH Director's New Innovator Awards are well-positioned to make significant – and potentially transformative – discoveries in a variety of areas.

Public–Private Partnerships (PPPs)

Another way the Roadmap encourages the creation of teams that cut across traditional boundaries is by fostering Public–Private Partnerships (PPPs). PPPs offer an opportunity to integrate several critical aspects of science capable of moving us into the future. Indeed, the vision of a personalized medicine – medicine that is able to promote health by personalized risk assessment and prevention, ameliorate disease by timely and effective interventions, and avoid toxic or morbid side effects of treatment by sensitive prediction – can only be promoted by combining the resources, insights, and tools of the public and private sectors.

Why partnerships? Scientific research is increasingly expensive, and PPPs allow for cost sharing across sectors. Science is increasingly complex, and PPPs permit early and substantive sharing of planning and implementation inclusive of scientists, clinicians, regulators, manufacturers, and the public. In this way, PPPs – properly

---

[5] NIH Roadmap for Medical Research, "Science News About Pioneer Awardees," http://nihroadmap.nih.gov/pioneer/AwardeeScienceNews.aspx

configured – facilitate faster and more efficient scientific work, thereby hastening the translation of discovery to benefits in public health.

Cost sharing can be understood in terms of leveraging the considerable public investment in research and promoting the effective translation of government-supported discovery to clinically available therapeutics and interventions. Examples of translation via technology transfer are numerous. They include the development of breast cancer drugs such as Taxol (whose use in this tragically common and important disease was developed in NIH's intramural research program) and the development of coated stents for use in angioplasty. The cost of commercialization and of meeting regulatory requirements was undertaken by private companies, thus making these benefits available to the public.

PPP: The Biomarkers Consortium

Another advantage to PPPs is that their multi-sector teams generate a unique form of synergy. Such synergy can offer novel solutions to vexing clinical problems, including the assessment of individual patients, the development of effective drugs, and the complications of navigating the regulatory review and approval process. The Biomarkers Consortium is a PPP that promises to generate just such synergistic solutions.

Biomarkers are measures of some aspect(s) of health or disease. "Biomarkers" comprise a wide array of biological indicators that can serve to identify risk, define diagnosis, stratify patients, predict outcomes, or signal response to therapy or progression of disease. Similarly, they can be measured by a wide variety of methods and platforms, including, for example, biochemical measures of protein or nucleic acid; images such as those collected from X-rays, MRIs or molecular imaging; and functional measures of immune system response or past exposures. Applying biomarkers in clinical practice or drug development requires the coordination of a wide range of basic, clinical, and regulatory scientific principles and processes. There must be a body of agreed upon standards for collection, measurement, and interpretation in specific contexts. A consortium including all such expertise, working within a coordinated and collaborative framework, facilitates the discovery, development and regulatory qualification of biomarkers.

The Biomarkers Consortium (BC) – one of the first partnerships supported by the Roadmap's PPP program – was established in October, 2006. The Foundation for NIH serves as the managing partner of this large and complex PPP; founding partners include NIH, FDA, and PhRMA; as well as CMS, BIO, numerous drug and biotechnology companies, academia, patient advocacy groups, and professional societies. Much care was devoted to developing a consortium structure and policies attentive to concerns about protecting the public health, working in a pre-competitive fashion, and developing public biomarker resources. Moreover, BC polices explicitly addressed ways to protect the public, and partners, from conflicts of interest and antitrust issues.

The BC's initial projects are currently under way; they include qualification studies relating to the use of FDG-PET imaging in non-small cell lung cancer and lymphoma. Projects evaluating the use of carotid MRI in atherosclerotic disease and in assessing the role of adiponectin as a marker of treatment response in diabetes are about to begin. Some 30 to 40 additional projects in areas of oncology, neuroscience, immunity and inflammation, and metabolic diseases, have been approved; still other specific project plans and associated agreements are being developed.

The promise of the BC lies not only in the anticipation of having a wider array of available qualified biomarkers for clinical and regulatory uses, but also in its role as a template for large multi-sector partnerships promoting cross-disciplinary science in a speedy and efficient manner. Partnerships are not likely to serve all purposes: basic discovery, for example, is not a reliable investment for industry, and its utility for specific diseases cannot always be predicted. For activities representing shared aims and goals, and where alignment of business principles and practices can be arrived at, PPPs offer a powerful approach to leveraging partners' investments through synergy.

The growing ties between academia and industry are leading to productive collaborations and innovations in healthcare. Academic stars routinely turn entrepreneurial, and private industries and venture capitalists constantly seek potentially valuable basic science breakthroughs worthy of support. These exciting trends also pose a serious challenge: how do we manage the conflicts of interest created when researchers develop economic stakes in their research? Conflicts of interest can undermine public confidence in scientific data and policy recommendations. They have also been shown to subtly influence even the most honest of people. For the Federal employees at NIH, we have adopted stringent restrictions on conflicts of interest. But for our grantees in universities and institutes around the country, we take a more nuanced approach.

The biomedical research community recognizes that some of the most knowledgeable experts in many fields have experience in both the public and private sectors. "Wearing both hats" may grant them unique insights into how to translate basic research into viable healthcare interventions. Further, one could argue that allowing great researchers to benefit financially from their own discoveries may not only be fair, but may also provide incentives to increase productivity. The risks of *excluding* more entrepreneurially inclined scientists from the research enterprise may well outweigh the risks of letting them continue to conduct research. However, the relative risk depends on how well the biomedical research enterprise can identify and manage conflicts. We are working diligently with scientists, universities, stakeholder groups, the Institute of Medicine, and policy makers to develop strategies to appropriately deal with conflicts of interest – to optimize the balance between maintaining the public trust and maximizing the public benefits of NIH research.

Tools

Of course, all the teamwork in the world would not take biomedicine very far if its teams weren't equipped with the tools needed to translate ideas into actions. In today's biomedical research labs, tools and technologies are being imported from an impressive spectrum of sciences, and finding new applications in the process. At the same time, biomedical needs are helping to shape new technologies. "Necessity" and "invention" play mutually reinforcing roles in the advance of today's biomedicine. The NIH Roadmap encourages this process by supporting the development of tools that will facilitate cross-cutting biomedical research and quicken the pace of its translation into clinical applications.

Nanomedicine

Nanomedicine offers a striking example of the Roadmap's support of research tool development. "Nano" refers to a unit of measure: one nanometer equals one billionth of a meter. It is the scale applied to atoms and molecules. "Nanomedicine" is an offshoot of nanotechnology, which is based on fundamental discoveries in physics and chemistry that now allow us to manipulate atoms and molecules in order to create nanomaterials. Because of their small size, nanomaterials often have quite different properties than do their larger counterparts. These novel properties have promising biomedical relevance. Over the past few years, NIH established eight Nanomedicine Development Centers across the U.S. These collaborative centers are staffed with multidisciplinary teams including biologists, chemists, material engineers, clinicians, mathematicians, and computer scientists. Breakthroughs in nanomedicine are already being achieved as experts team up to devise novel nanomaterials that aid in drug delivery, serve as tissue scaffolds, or enhance imaging scans in patients.

Structural Biology

A healthy mind and body require the coordinated action of billions of proteins – molecular workers that build our cells and even allow us to think, smell, eat, and breathe. Proteins have unique three-dimensional shapes that allow them to accomplish their particular tasks. A protein's shape is so essential to its ability to function properly that a structural error in even one protein can have major health consequences. How, then, can medicine help ensure that such errors do not occur?

Answering this question has proven to be as difficult as it is critical. Efforts to study the structures of protein membranes have been only occasionally successful. A limiting step in determining protein structures is our inability to produce purified samples of the proteins in quantities sufficient for analysis. Proteins that are tightly bound to the membranes of our cells have been the most difficult to study. Yet, membrane proteins not only account for about 30% of the proteins in a cell, but they also turn out to be one of the most important classes of proteins for health. Moreover, they are major drug targets for the development of disease treatments. These proteins were therefore chosen for special Roadmap attention.

The Structural Biology Roadmap program is a strategic effort to create a "picture" gallery of the molecular shapes of proteins in the body. This will require the development of rapid, efficient, and dependable methods to produce protein samples that scientists can use to determine the three-dimensional structure of a protein. Once developed, these methods will streamline and systematize research efforts, producing a routine that will help researchers clarify the role of protein shape in health and disease.

During the first phase of the Structural Biology Roadmap (FY2004–2008), the NIH Roadmap funded two Centers for Innovation in Membrane Protein Production that enabled interdisciplinary groups of scientists to develop innovative methods for producing large quantities of membrane proteins. In addition, a number of small exploratory (R21) and regular research grants (R01) were awarded to individual investigators to broaden the base of innovative ideas under development.

These investments have already produced considerable advances in methods and in solved structures. Most notably, researchers determined the structure of the beta-2 adrenergic receptor. This receptor, which is adrenaline's target, is also the target of numerous drugs. Moreover, it is a prime example of a large family of important cell regulatory molecules known as G-protein coupled receptors (GPCRs). GPCRs mediate our interactions with the world outside our bodies by detecting sensory perceptions such as light and taste; they are also essential to the maintenance of our internal environment, acting as relays for molecules such as neurotransmitters and hormones. So significant was this discovery that *Science* magazine listed it among its top ten breakthroughs for 2007. [6]

Thus far, however, most work has been done on relatively *simple* membrane proteins. Many membrane proteins are complex biological machines consisting of several interlocking proteins working together. To understand how these machines work – and to learn how to fix them when they don't – researchers need to view the protein complexes in several different orientations, mimicking the way these assemblies twist and bend inside living cells. NIH anticipates that scientists will require about a decade of intense work to achieve the project's most ambitious goal: the ability to routinely predict the shape and action of a biological machine from its DNA script. The next phase of the NIH Structural Biology Roadmap (FY2009–2013) will be devoted to developing the means necessary to achieve this goal.

National Technology Centers for Networks and Pathways

In the human body, all biological components – from individual genes to entire organs – work together to promote normal development and sustain health. This amazing feat of biological teamwork is made possible by an array of intricate and interconnected pathways that facilitate communication among genes, molecules, and cells.

Limitations of proteomics technologies often force investigators to treat dynamic systems as either static or as binary options between static states. As with early photography, current approaches to proteomics involve long exposures that capture broadly defined "images" such as "normal vs. diseased," "the yeast interactome," or "the nuclear pore complex." We are largely blind to the dynamics of systems which we know are not static but which must be treated as such at present because of

---

[6] "Breakthrough of the year: the runners-up," *Science* 318 (2007): 1844–1849.

inadequate tools. Transient interactions, rapid changes in protein activity or location, and post-translational modifications control critical regulatory steps in biology, yet they are like a bird flying through the frame of a carefully composed long exposure: invisible.

New strategies complementary to conventional proteomics are necessary to help us determine the rapid, dynamic changes that control physiology. The National Technology Centers for Networks and Pathways (TCNPs) create technologies to measure the dynamics of protein interactions, modifications, translocation, expression, and activity, and to do so with temporal, spatial, and quantitative resolution. The program is intended to bridge the quantitation and interaction aspects of proteomics, breaking out of the artificially static view of complex systems. [7] Five independent centers cooperate in a networked national effort to develop instrumentation, biophysical methods, reagents, and infrastructure for temporal and spatial characterization of complex biochemical pathways and networks of protein interactions. The centers are also tasked with providing broad access to the technologies, methods, and reagents they develop, as well as providing appropriate interdisciplinary academic and peer training for biomedical researchers.

National Centers for Biomedical Computing

Clearly, biomedical research is rapidly moving beyond the microscopes, test tubes, and Petri dishes that have been its defining tools. Sophisticated techniques adapted from physics, chemistry, and engineering enable scientists to use computers and robots to separate molecules in solution, read genetic information, reveal the three-dimensional shapes of natural molecules, and take pictures of the brain in action. All of these techniques generate large amounts of data, and there is no way to manage these data by hand. Biology is fast changing into a science of information management. What researchers need are computer programs and other tools to evaluate, combine, and visualize their voluminous data.

The NIH Roadmap program called the National Centers for Biomedical Computing was created to generate the software and data management tools to serve as fundamental building blocks for 21st century medical research. In this program, "big science" and "small science" work hand-in-hand to develop a system that will ultimately resemble the integrated software packages for office tools installed on most home computers today. This system will allow information to be traded seamlessly and cooperatively analyzed. It will allow our best minds to work together effectively to tackle unsolved mysteries – such as the role of heredity in individuals' different responses to medicines and the complex interplay of genetic and environmental factors in common diseases such as Alzheimer's disease, heart disease, cancer, and diabetes.

---

[7] Douglas M. Sheeley, Joseph J. Breen, and Susan E. Old, "Building integrated approaches for the proteomics of complex, dynamic systems: NIH programs in technology and infrastructure development," *Journal of Proteome Research* 4 (2005): 1114–1122.

Patient Translations

The NIH's mission extends beyond the laboratory – even beyond laboratories boasting the latest technology and the most integrated approaches to science – and to the improvement of human health. In this still-young century, scientific discoveries have blended with the recent doubling of the NIH budget, justly raising public expectations for rapid progress in fulfilling this mission. Clinical research is a vital component of progress toward improving America's health. But while clinical research helps ensure that new treatments are safe and effective, it is a lengthy and sometimes inefficient process. Growing barriers between clinical and basic research, along with the ever-increasing complexities of conducting clinical research, are making it more difficult to translate new knowledge to the clinic – and back again to the bench. These challenges are limiting professional interest in the field and hampering the clinical research enterprise at a time when it should be expanding. The Roadmap supports an array of programs that aim to accelerate and strengthen clinical research by adopting a systematic infrastructure – from tools and training to discipline-building – that will better serve the evolving field of scientific discovery.

RAID and PROMIS

One roadblock to the safe and effective production of new therapeutic interventions lies in the limited availability of key resources. Two Roadmap initiatives – the NIH Rapid Access to Interventional Development (RAID) Pilot Program and the Patient-Reported Outcomes Measurement Information System (PROMIS) – have been devised to increase the availability of two key resources: funding and related support, and a uniform way of understanding and assessing patients' symptoms.

RAID provides funding and support for the development of novel therapeutic interventions for the treatment of uncommon disorders. While the translation of such interventions is sometimes facilitated by public–private partnerships, high-risk ideas or therapies for uncommon disorders frequently do not attract private sector investment. Where private sector capacity is limited or not available, public resources can bridge the gap between discovery and clinical testing so that more efficient translation of promising discoveries may take place. RAID is a pilot program that will make available, on a competitive basis, certain critical resources needed for the development of new small molecule therapeutic agents. Projects in both the early and late stages of pre-clinical development are suitable for NIH-RAID applications.

The PROMIS initiative is helping to overcome our limited ability to assess the symptoms that patients experience in response to therapeutic interventions for a wide array of disorders. Patient-reported outcomes (PROs), such as pain, fatigue, physical functioning, emotional distress, and social role participation, have a major impact on quality of life. Clinical measures of health outcomes, such as x-rays and lab tests, may have minimal relevance to the day-to-day functioning of patients with chronic diseases. Often, the best way patients can judge the effectiveness of treatments is by changes in symptoms. The goal of PROMIS is to improve the reporting and quantification of changes in PROs by developing a rigorously tested measurement tool that uses recent advances in information technology, psychometrics, and qualitative, cognitive, and health survey research. In the process, the initiative is creating new paradigms for how clinical research information is collected, used, and reported.

Clinical and Translational Science Award Program

Aimed at generating new drugs, devices, and treatment options, the NIH "bench to bedside to community" enterprise is designed to break down the historic walls separating basic scientists, clinical researchers, and community practitioners. The reengineered home for these translational teams is supported through the Clinical and Translational Science Award (CTSA) program. Led by NIH's National Center for Research Resources, the CTSA program creates a definable academic home for the emerging discipline of clinical and translational science at institutions across the country. To create this home, the program encourages local flexibility, allowing each participating institution to determine whether to establish a center, department, or institute in clinical and translational science. In its first year, 2006, the program made awards to 12 academic health centers. When fully implemented in 2012, approximately 60 institutions will be linked together to energize the discipline of clinical and translational science. [8]

The members of this CTSA consortium are expected to serve as a magnet that attracts basic, translational, and clinical investigators, community clinicians, clinical practices, networks, professional societies, and industry to facilitate the development of new professional programs and research projects. We anticipate that these new institutional arrangements, coupled with innovative advanced degree programs, will foster the development of a new discipline of clinical and translational science – one that will be much broader and deeper than the classical domains of translational research and clinical investigation have been on their own. [9]

Expanding the Roadmap

The NIH Roadmap is intended to act as an *incubator* for innovative and interdisciplinary research. As such, it provides this research with an opportunity to grow, thrive and, eventually, to lead an independent existence. Initiatives are constantly reviewed and, over time, will be cycled out of the Roadmap. Consequently, NIH is continuously evaluating new directions for future initiatives. We recently added the Human Microbiome Project, which uses genomic technologies to illuminate the role of our resident microbes in health and disease. We are also funding a new initiative in epigenomics. Whereas epigenetics focuses on processes that regulate how and when certain genes are turned on and turned off, epigenomics studies epigenetic changes across many genes in a cell or entire organism. The epigenomics initiative

---

[8] Clinical and Translational Science Awards Consortium, http://www.ctsaweb.org/

[9] Siri Carpenter, "Carving a career in translational research," *Science* 317 (2007): 966–967.

will build upon our knowledge of the human genome and help us better understand the role of the environment in regulating genes that protect our health or make us more susceptible to disease.

From Crossroads to Future Horizons

The NIH Roadmap for Medical Research was conceived in response to the challenges and opportunities of our time; and it has been forged by the ideals of integrated science. The broad outlines of its destination are evident. For biomedicine in particular, however, the Roadmap's ultimate destination is nothing short of effecting a new paradigm for medical practice.

Ever since the days of Hippocrates, medical practice has been driven by a goal of *curing* disease: symptoms develop, and a person, who has become a patient, seeks medical intervention to cure those symptoms. (If one counts the intervention of deities, the curative paradigm far predates the Hippocratic physicians.) Present-day scientific advances suggest that we can – and should – aspire to a new medical paradigm. In this paradigm, medical practitioners will use their deep understanding of living systems and disease processes to *predict* the course of disease in individual patients before symptoms ever strike. These practitioners will then *personalize* their (pre-symptomatic) interventions to fit the particular genetic, social, and environmental needs of each patient. This approach will only work if patients actively *participate* in their own healthcare – and if communities link people to clinical trials and medical institutions in a network of mutual support. Ultimately, this predictive, personalized, and participatory approach will usher in a new era of medical care: an era in which our knowledge will give us the power to *preempt* symptoms before they ever transform people into "patients." In other words, these "4 Ps" will bring about a shift from a *curative* to a *preemptive* medical paradigm. An integrated approach to science will be a fundamental driver of this shift.

I believe these advances in the life sciences will have applications that extend even beyond the improvement of human health. They will have significant applications to the interrelated ailments – from crop growth and energy supply to global warming – that threaten our planet. We must do all we can to pave the way for these advances – and we must do it now. If we succeed, those future generations of historians will look back upon us as sage stewards of our planet. Seen in this light, our failure cannot be contemplated.

James J. Duderstadt, President Emeritus and University Professor of Science and Engineering at the University of Michigan

Dr. James J. Duderstadt served as the President of the University of Michigan from 1988 to 1996. He currently chairs a number of national commissions related to federal science policy, higher education, information technology, and energy sciences and holds a university-wide faculty appointment as University Professor of Science and Engineering at the University of Michigan and as Director of the Millennium Project, a research center exploring the impact of over-the-horizon technologies on society.

Many have suggested that academic science, industry, and government should work together to find solutions to broad social problems. Dr. Duderstadt, how would you describe recent trends of convergence between the efforts of the academy, federal and state government, and industry?

The efforts of universities and faculty members to capture and exploit the soaring commercial value of the intellectual property created by research and instructional activities create many opportunities and challenges for higher education. Clearly there are substantial financial benefits to those institutions and faculty members who strike it rich with tech transfer. In the 1980s, it was the "red Ferrari in the parking lot" syndrome, as the first signs of faculty wealth from tech transfer began to appear. In the booming days of the dot-coms, the more typical story is of the young assistant professor of computer science telling his department chair, "I'm going to take a one year leave of absence to start up a company. If I'm successful, I probably won't return, but at least you may get a million dollar gift out of me. If I'm not successful, then I'll return and see if I can get tenure." Or yet another faculty member, who informs his chair that he has set up a small foundation financed by his recent IPO, apologizing that his first gift will be only $10 million, but he expects his contributions to rise rapidly.

Each of these stories is true (although the Ferrari belonged to the wife of a professor who had struck it rich from a best selling textbook). But there are also many signs that the commercialization of intellectual property has its downside as well. Today scientists sign agreements requiring them to keep both the methods and the results of their work secret for a certain period of time. More than a quarter of US geneticists say they can't replicate published findings because other investigators will not give them relevant data or materials. There is growing evidence suggesting that industrial sponsorship actually influences the outcome of scientific work.[10] Universities are encountering an increasing number of conflict of interest cases, stimulated by the exploding commercial value of intellectual property and threatening not only institutional integrity but even human life in conflicted clinical trials.

In recent years, many universities seem to have adopted the attitude that "What is good for General Motors – or rather, consistent with the Bayh-Dole Act – is good for the country."[11] They recognize and exploit the increasing commercial value of the intellectual property developed on the campuses as an important part of their mission (and part of their reward as well, I might add.) This has infected the research university with the profit objectives of a business, as both institutions and individual faculty members attempt to profit from the commercial value of the products of their research and instructional activities. Universities have adopted aggressive commercialization policies and invested heavily in technology transfer offices to encourage the development and ownership of intellectual property rather than its traditional open sharing with the broader scholarly community. They have hired teams of lawyers to defend their ownership of the intellectual property derived from their research and instruction. On occasions, some institutions and faculty members have set aside the most fundamental values of the university, such as openness, academic

---

[10] "Data Hoarding Blocks Progress in Genetics", *Science*, Vol 295, January 25, 2002, p. 599.

[11] Editor's Note: Enacted on December 12, 1980, the Bayh-Dole Act (P.L. 96-517, Patent and Trademark Act Amendments of 1980) created a uniform patent policy among the many federal agencies that fund research, enabling small businesses and non-profit organizations, including universities, to retain title to inventions made under federally-funded research programs. This legislation was co-sponsored by Senators Birch Bayh of Indiana and Robert Dole of Kansas.

freedom, and a willingness to challenge the status quo in order to accommodate this growing commercial role of the research university.[12]

But what is the public interest here? As Donald Kennedy has noted, " 'Public interest' has two translations. In the more technical, political science sense, it refers to those attributes of a venture or an organization that supports the larger society, benefiting the welfare of all the people. More colloquially, it can also mean what the public cares about, what it is interested in."[13]

Do you see an increase in the commercial activities of academic institutions, at either the federal or state level?

The Association of University Technology Managers[14] estimate that during FY2000 universities and their faculties collected more than $1 billion in royalties, created 368 spin-off companies, filed for 8,534 patents, and executed 3,606 licenses and options. While this royalty figure is some 40 percent higher than in FY1999, it includes several one-time events such as $200 million paid by Genetech to UCSF to settle a patent dispute and several universities cashing in their equity interest from earlier spin-off activities. Furthermore, it is also true while some universities benefited greatly from these commercial activities, most received less than $1 million in royalties, which was frequently not even sufficient to cover the costs of their technology transfer activities. Actually, from the earliest days of the Bayh-Dole Act of 1980, only a few inventions and discoveries have struck it rich for universities (e.g., recombinant DNA at UCSF and Stanford, Lycos at Carnegie-Mellon, carboplatin at Michigan State, and, of course, Gatorade at the University of Florida). In contrast, many individual faculty members have benefited considerably from equity interest in spin-off companies through IPOs and other financial events as my anecdotes in the introduction suggest.

At the level of the states, governments are sending public research universities clear signals to commercialize their discoveries in an effort to stimulate local economic development.[15] Nearly one-third of the governors have called on legislatures to pump money into campus research and tech transfer programs.[16] Several states

---

[12] Eyal Press and Jennifer Washburn, "The Kept University", *Atlantic Monthly*, March, 2000, pp. 39–54.

[13] Donald Kennedy, *Academic Duty* (Cambridge: Harvard University Press, 1999).

[14] *Annual Survey of Technology Licensing Activity*, FY2000, Association of University Technology Managers; see also Goldie Blumenstyk, "Income from University Licenses on Patents Exceeded $1 Billion", *The Chronicle of Higher Education*, March 22, 2002.

[15] Peter Schmidt, "States Push Public Universities to Commercialize Research" *The Chronicle of Higher Education*, March 29, 2002.

[16] Here it is worth noting that my own state, Michigan, committed $50 million per year from their tobacco settlement payments to support biomedical research in a "Life Sciences Corridor" stretching from Detroit to Grand Rapids. However, it is also worth noting that Michigan was one of only three states choosing not to deploy any of the tobacco funds for their stated intent, to ameliorate teenage smoking.

have changed their laws to eliminate barriers to public-private collaboration, including giving for-profit companies unprecedented access to public university research facilities, while encouraging public universities and their employees to hold a financial stake in companies. Even conflict of interest and freedom-of-information laws have been throttled back to protect proprietary activities in nearly half of the states.

What about the effect of the market in influencing the direction of the university?

Today our society is evolving rapidly into a post-industrial, knowledge-based society, a shift in culture and technology as profound as the social transformation that took place a century ago as an agrarian America evolved into an industrial nation.[17] Industrial production is steadily shifting from material- and labor-intensive products and processes to knowledge-intensive products and services. A radically new system for creating wealth has evolved that depends upon the creation and application of new knowledge.

In a very real sense, we are entering a new age, an age of knowledge, in which the key strategic resource necessary for prosperity has become knowledge itself, that is, educated people and their ideas. Unlike natural resources, such as iron and oil, that have driven earlier economic transformations, knowledge is inexhaustible. The more it is used, the more it multiplies and expands. But knowledge is not available to all. It can be absorbed and applied only by the educated mind. Hence as our society becomes ever more knowledge-intensive, it becomes ever more dependent upon those social institutions such as the university that create knowledge, that educate people, and that provide them with knowledge and learning resources throughout their lives.[18]

This increasing economic value of the university and its products, along with other factors such as changing social needs, economic realities, and rapidly advancing technology, have created powerful market forces acting upon and within higher education. Even within the traditional higher education enterprise, there is a sense that the arms race is escalating, as institutions compete ever more aggressively for better students, better faculty, government grants, private gifts, prestige, winning athletic programs, and commercial market dominance. Faculty members, as the key sources of intellectual content in both instruction and research, increasingly view themselves as independent contractors and entrepreneurs, seeking ownership and personal financial gain.

With the emergence of new competitive forces and the weakening influence of traditional regulations, the higher education enterprise is entering a period of restructuring similar to that experienced by other economic sectors such as health care, communications, and energy. Higher education is breaking loose from the

---

[17] Peter F. Drucker, "The Age of Social Transformation," *Atlantic Monthly*, November 1994, 53–80; Peter Drucker, "The Next Society: A Survey of the Near Future," *The Economist*. (3 November 2001) 356(32): 3–20.

[18] Derek Bok, *Universities and the Future of America* (Durham: Duke University Press, 1990).

moorings of physical campuses, even as its credentialing monopoly begins to erode. It appears to be evolving from a loosely federated system of colleges and universities serving traditional students to, in effect, a global knowledge and learning industry driven by strong market forces.

As our society becomes ever more dependent upon new knowledge and educated people, upon knowledge workers, this global knowledge business must be viewed clearly as one of the most active growth industries of our times. Today it is estimated that higher education represents roughly $225 billion of the $665 billion education market in the United States.[19] But even these markets are dwarfed by the size of the "knowledge and learning" marketplace, a convergence of education, communications, information technology, and entertainment sectors, estimated in excess of $2 trillion.

This perspective of a market-driven restructuring of higher education as an industry, while perhaps both alien and distasteful to the academy, is nevertheless an important framework for considering the future of the university. These social, economic, technological, and market forces are far more powerful than many within the higher education establishment realize. They are driving change at an unprecedented pace, perhaps even beyond the capacity of our colleges and universities to adapt. There are increasing signs that our current paradigms for higher education, the nature of our academic programs, the organizations of our colleges and universities, the way that we finance, conduct, and distribute the services of higher education, may not be able to adapt to the demands and realities of our times.

As each wave of transformation sweeps through our economy and our society, with an ever more rapid tempo, the existing infrastructure of educational institutions, programs, and policies becomes more outdated and perhaps even obsolete. It is clear that no one, no institution, and no government, will be in control of the emergence and growth of the knowledge industry. It will respond to forces of the marketplace. And perhaps this is the most serious threat of the emerging competitive marketplace for knowledge and learning: the danger that it will not only distort but erode the most important values and purposes of the university. In a highly competitive market economy, short-term commercial opportunity and challenges usually win out over long-term public interests.

Given the potential positive outcomes associated with such industry-driven research, can you enumerate any concerns?

In the past, the public purposes of our universities were determined primarily by public policy and public investment. Today the marketplace may be redefining these roles. The ties between universities and the corporate world have proliferated and changed over recent decades. There has been a shift in the priorities of the university, away from the pursuit of knowledge and the education of the next generation and instead toward responding to the commercial lure of the marketplace.

---

[19] Michael T. Moe, *The Knowledge Web* (Merrill-Lynch, New York, 2000).

While partnerships between universities and industry have existed for many years, in the past they tended to rely on traditional relationships such as the hiring of graduates, the use of faculty consultants, or the sponsorship of research. Financial associations with private industry were largely confined to companies awarding grants to academic institutions for research in areas of mutual interest. Companies played no part in designing or analyzing the studies; they did not house the data, and they certainly did not write the papers and control the publications of results.

Things have changed dramatically in the past decade. Arm's length relationships are a thing of the past, and financial arrangements go far beyond simple grant support. In some research universities, the conflict of interest policies have been designed primarily to comply with federally funded research, while the increasing flow of privately funded research is eroding university-wide compliance with the spirit and letter of the federal guidelines. New forms of hybrid institutions have emerged to facilitate joint industry-university collaborations that are not formally covered by faculty policies. The increasing trend for students at the graduate and undergraduate level to be involved in proprietary work with sponsoring corporations can create conflicts for which most university government committees have few policies and sometimes no oversight.

Of particular concern is the attention paid within the university research community to the commercialization of technology and discoveries, sometimes with the potential for very large financial rewards to individual faculty members under prevailing technology transfer policies and practices. The traditional belief of universities that proprietary claims were fundamentally at odds with their obligation to disseminate knowledge as broadly as possible fell by the wayside with the Bayh-Dole Act of 1980. This legislation obliged those receiving federal funds for research to make strong efforts to promote the commercialization of their discoveries. From that time forward, faculty researchers were expected to be aware of the potential commercial value of their work and their institutions were obliged to create the infrastructure that would facilitate patenting, marketing, and licensing their faculty's discoveries. It didn't take long for universities to realize that the Bayh-Dole mandate had the potential for becoming a "cash cow" for the institution and the faculty. Universities invested heavily in technology transfer and licensing offices with the missions of developing, protecting, and marketing of intellectual properties.

Did the federal government favor industry-driven research at academic institutions, and if so, how?

The federal government played a major role in stimulating and sustaining the American research university through the government-university research partnership first articulated in Vannevar Bush's *Science, the Endless Frontier* report.[20]

---

[20] Vannevar Bush, *Science–The Endless Frontier* (National Science Foundation, Washington, 1945).

It has similarly triggered the explosion in campus activities designed to capture and exploit the commercial value of the intellectual property created by federally sponsored research through federal policies such as the Bayh-Dole Act of 1980. This legislation allows universities to retain the ownership of commercially valuable intellectual property produced in government-sponsored research. Universities have responded by providing strong incentives to their faculty and creating technology transfer offices to identify, protect, patent, license, and spin-off commercially valuable products and companies. As one data point, prior to the Bayh-Dole Act of 1980, universities produced roughly 250 patents a year (most of which were never commercialized). In 2000, universities filed for 8,534 patents and spun off 368 companies.

Prior to the Bayh-Dole Act, technology transfer occurred primarily through publication in scientific journals, technical consulting, continuing education and extension services, and the employment of trained graduates. To this array, Bayh-Dole added the transfer of a property right as the result of ownership of the intellectual property generated during the conduct of research, as manifested by patents, copyrights, trademarks, trade secrets, or a proprietary right in the tangible products of research. Fundamental to Bayh-Dole was the certainty that if the universities were the owners of inventions from research, they could grant exclusive licenses stimulating the private sector to invest in development.

The underlying tenant of the Bayh-Dole Act is that inventions resulting from federally funded research should be owned by universities and provided to exclusive licenses to industry for commercial development in the public interest. The act was based on the belief that a non-exclusive licensing policy simply is not effective in technology transfer. It is the incentive inherent in the right to exclude conferred upon the private owner of a patent that is the inducement to development efforts necessary to the marketing of new product. What is available to everyone is of interest to no one. Proponents of Bayh-Dole note that when the government held title to inventions under the policy that the inventions should be available to all, much the same as if the invention had been disclosed in a publication, the patent system could not operate in the manner in which it was intended.

But is this true? Although the recent increases in university patenting and licensing are widely assumed to be the direct consequences of Bayh-Dole, empirical evidence suggests that the impact of this activity on the content of academic research has been modest.[21] The growing importance of biomedical research, much of which relied on federal support that expanded significantly during the 1970s, was at least as important as Bayh-Dole in explaining increased university patenting and licensing after 1980. Other factors also encouraged the growth of university patenting in this and other areas such as judicial decisions that declared that "engineering molecules" were patentable. It seems clear that an array of developments in research,

---

[21] David C. Mowery, Richard R. Nelson, Bhaven N. Sampat, and Arvids A. Ziedonis, "The Effects of the Bayh-Dole Act on U.S. University Research and Technology Transfer", in *Industrializing Knowledge*, ed. L. Branscomb and R. Florida (MIT Press, Cambridge, 1999).

technology, industry, and policy combined to increase US universities in technology licensing, and Bayh-Dole, while important, was not determinative.

Furthermore, the Bayh-Dole Act represents an application of the "linear model" to science and technology policy, assuming that if basic research results can be purchased by would-be developers, thereby establishing clear potential for the commercial development of these results, commercial innovation will be accelerated. The earlier concept of a linear progression of basic research to applied research to commercial development to marketable products, a fundamental assumption of the *Science, the Endless Frontier*[22] policies that have governed university research for the past half century, has been replaced by a nonlinear process in which basic and applied research, development and commercialization are mixed, a la Pasteur's Quadrant[23] (or Jeffersonian science).

The theory behind Bayh-Dole (that companies need exclusive patent rights to pick up, develop, and commercialize the results of university research) seems in conflict with the fact that patents tend to restrict use of scientific and technological information and open publication, which facilitates wider use and application of such inventions and knowledge. Are patents or restrictive licenses really necessary to achieve application? Should such licenses be negotiated by universities, institutions not known for their commercial expertise? Does the presence of a university-assigned patent and the requirement for licensing delay and narrow technology transfer? There is as yet little empirical evidence in support of this principle.

There are still other challenges to the conventional Bayh-Dole doctrine. Are universities' patenting efforts increasing or reducing the social returns to the results of the publicly funded research performed on their campuses? Are universities' expanded efforts to patent inputs into the scientific research process impeding progress? It is certainly true that with the increasing emphasis on disclosure, patenting, and licensing, more of what universities naturally would have produced and placed in the public domain now is subject to more complex administrative procedures. These policies may raise the costs of use of these research results in both academic and non-academic settings, as well as limiting the diffusion of these results.

Donald Kennedy made an excellent further point in a recent editorial in *Science*. He suggests that just as Vannevar Bush's *Endless Frontier* changed fundamental science from a venture dependent on small privileged elites into a vast publicly owned enterprise, Bayh-Dole and related federal policies is driving university research toward the private sector, fueled by the mobilization of philanthropy and corporate risk capital. Continuing the frontier motif, he suggests we might regard the current framework characterizing technology transfer as the "Great Enclosure." Just as the Homestead Act of 1862 transformed the American frontier from public land into a checkerboard of individually owned holdings by allocating land virtually free to

---

[22] Vannevar Bush, *Science–The Endless Frontier* (National Science Foundation, Washington, 1945).

[23] Donald E. Stokes, *Pasteur's Quadrant: Basic Science and Technological Innovation* (Brookings Institution Press, Washington, 1997).

those who would promise to live on and improve it, the largely public domain of basic research is now moving into private hands by yet another federal act, Bayh-Dole, that allows universities or individual scientists to claim ownership of the intellectual property created by federally sponsored research. Interestingly, these enclosure revolutions came about in the same way: both were implemented by purposeful government intervention, accomplished through statute.

Kennedy contends that while this has brought some major benefits, it has also been accompanied by significant costs. New problems of conflict of interest, royalty distribution, and the propriety of commercial relationships have arisen for faculty members and university administrators alike. The contemporary enclosure of the Endless Frontier is replicating the history of the Homestead Act, yielding patent disputes, hostile encounters between public and private ventures, and faculty distress over corporate deals with their universities. Sometimes government action is unintended, such as the recent Executive Order on stem cell research that promises to transform a major public program into the private sector. Many observers, noting these costs, advocate policies for reversing privatization.

Would you suggest that these trends may ultimately lead to a major change in the fundamental mission of the academy?

Transferring university-developed knowledge to the private sector fulfills a goal of federally funded research by bringing the fruits of research to the benefit of society. With this important technology transfer comes increasingly close relationships between industry and universities. While this provides benefits to society, it also increases the risk of academic research being compromised by constraining open publication of research methods and results while diverting faculty from more fundamental research topics not so directly linked to commercial outcomes. Ironically, it has been the freedom of universities from market constraints that is precisely what allowed them in the past to nurture the kind of open-ended basic research that led to some of the most important (and least expected) discoveries in history.

There remains considerable uncertainty concerning just how universities should approach the commercialization of the intellectual property associated with campus-based research and instruction. Beyond the traditional triad of teaching, research, and service (or in more contemporary language, learning, discovery, and engagement),[24] it is useful to consider the "products" of the university as educated people, content, and knowledge services. Yet content, that is intellectual property, cannot be bottled and marketed like other commercial products. It exists in the minds of people, the faculty, staff, and students of the university. As such, it can simply walk out the door.

---

[24] *Kellogg Commission on the Future of State and Land-Grant Universities*, Renewing the Covenant. (National Association of State Universities and Land-Grant Colleges, Washington, 2000).

So how do universities handle content? Traditionally they have used the library model, that is, they distribute knowledge freely through open publication (and then, occasionally, are forced to buy it back in the form of expensive journals from commercial publishers). In the wake of Bayh-Dole, they have swung to the other extreme by attempting to capture, patent, and license the intellectual property resulting from their scholarly and instructional activities, relying on armies of lawyers to defend this ownership (much as the NCAA attempts to capture and control all of the riches generated by college sports). The past two decades have seen technology transfer shift from the "library" toward the "NCAA" model, in which private profit has become a stronger motivating force than public interest.

Of course, although the federal government has encouraged and facilitated this shift through policies such as Bayh-Dole, it certainly does not require it. Indeed, the National Institutes of Health state quite clearly that "Universities have no duty to return value to shareholders, and their principal obligation under the Bayh-Dole Act is to promote utilization, not to maximize financial returns. It hardly seems consistent with the purposes of the Bayh-Dole Act to impose proprietary restrictions on research tools that would be widely utilized if freely disseminated." Furthermore, while disclosure, patenting, and licensing intellectual property may be appropriate for some areas such as the product-orientation of biomedical research, it may not be an effective mechanism for very rapidly evolving areas such as information technology or instructional content.

Let me suggest an even bolder approach. Suppose that in return for strong public support, the nation's public universities could be persuaded to regard all intellectual property developed on the campus through research ... as in the public domain. They could encourage their faculty to work closely with commercial interests to enable these knowledge resources to serve society, without direct control or financial benefit to the university, perhaps by setting up a "commons" environment adjacent to the campus (either geographically or virtually) where technology transfer was the primary mission. This might be just as effective a system for transferring technology as the current Bayh-Dole environment for many areas of research and instruction. Furthermore, such an unconstrained distribution of the knowledge produced on campuses into the public domain seems more closely aligned with the century old spirit of the land-grant university movement.

What kind of changes do you believe are necessary to balance the goals of the University and the demands of industry in terms of technology transfer?

It is important in such discussions to always keep in mind the fundamental purposes and values of the university. In preparing for this discussion, I read back over the technology transfer policies of the University of Michigan, which begin with the statement: "The mission of the University is to generate and disseminate knowledge in the public interest. Essential to this mission are two fundamental principles: open scholarly exchange and academic freedom." And, of course, this is the issue in a nutshell: the degree to which the increasing commercialization of the academy is threatening its most fundamental mission and values.

As Henry Rosovsky put it at a recent meeting of American and European educators, the marriage between universities and industry is "against nature."[25] It represents a symbiotic relationship, between two unlike organisms with vastly different characteristics and objectives. The values of the university involve freedom of inquiry, the open sharing of knowledge, a commitment to rigorous study, and a love of learning. The goals of the marketplace are return on investment and shareholder value.

What is the public interest in the transfer of knowledge from the campus to society through commercial avenues? How are the rules and expectations characterizing the interaction between the university and the commercial marketplace changing? Is there an appropriate balance of public and private interests in today's universities? How are policies, practices, and dialog concerning the relationship between the university and industry affecting the traditional scholarly mission and sense of community on the campuses? Do universities and faculty have the necessary tools to manage the complexity of new relationships with industry? These are the questions that remain before us, and these are the issues that should be addressed through further dialog both on the campuses and with those served by the university.

The market forces driven by the increasing commercial value of the knowledge produced on our campuses are powerful indeed. Yet, if they are allowed to dominate and reshape the higher education enterprise without constraint, some of the most important values and traditions of the university will likely fall by the wayside. Will higher education retain its special role and responsibilities, its privileged position in our society? Will it continue to prepare young students for roles as responsible citizens? Will it provide social mobility through access to education? Will it challenge our society in the pursuit of truth and openness? Or will it become, both in perception and reality, just another interest group driven along by market forces? As we assess these market-driven emerging learning structures, we must bear in mind the importance of preserving the ability of the university to serve a broader public purpose.

The American university has been seen as an important social institution, created by, supported by, and accountable to society at large. The key social principle sustaining the university has been the perception of education as a public good – that is, the university was established to benefit all of society. Like other institutions such as parks and police, it was felt that individual choice alone would not sustain an institution serving the broad range of society's education needs. Hence public policy dictated that the university merited broad support by all of society, rather than just by the individuals benefiting from its particular educational programs.

Yet, today, even as the needs of our society for post-secondary education intensifies, we also find an erosion in the perception of education as a public good

[25] Henry Rosovsky, "And the Walls Come Tumbling Down" in *The Glion III Symposium*, ed. by Luc Weber and Werner Hirsch, (Paris: Economica, 2002).

deserving of strong societal support.[26] State and federal programs have shifted from investment in the higher education enterprise (appropriations to institutions or students) to investment in the marketplace for higher education services (tax benefits to students and parents). Whether a deliberate or involuntary response to the tightening constraints and changing priorities for public funds, the new message is that education has become a private good that should be paid for by the individuals who benefit most directly, the students. Government policies that not only enable but intensify the capacity of universities to capture and market the commercial value of the intellectual products of research and instruction represent additional steps down this slippery slope.

Education and scholarship are the primary functions of a university, its primary contributions to society, and the most significant roles of the faculty. When universities become overly distracted by other activities, they not only compromise these core missions but they also erode their priorities within our society. The shifting perspective of higher education from that of a social institution, shaped by the values and priorities of broader society, to, in effect, an industry, increasingly responsive to the marketplace only intensifies this concern. While it is important that the university accept its responsibility to transfer the knowledge produced on its campus to serve society, it should do so in such a way as to preserve its core missions, characteristics, and values. In particular, the nature of higher education as a public good rather than simply a market commodity needs to be recognized by higher education and re-established by strong public policy and public investment both at the federal level and at the level of our states and communities, since the future of the university in an ever more knowledge-driven society is clearly a national concern.

---

[26] Robert Zemsky, "Rumbling," Policy Perspectives, Pew Higher Education Roundtable, sponsored by the *Pew Charitable Trusts* (Philadelphia: Institute for Research on Higher Education, April 1997).

Part II
The Integration of Academic Science and Industry

Chapter 3
Should Business and Industry Create Integrative Partnerships with Academic Science?

Stanley Aronowitz, Distinguished Professor of Sociology and Director of the Cultural Studies Program at the Graduate Center, City University of New York

Dr. Stanley Aronowitz has taught at the Graduate Center of the City University of New York since 1983, where he is Distinguished Professor of Sociology and Urban Education. Author of twenty-five books on topics relating to labor, social

M.E. Brint et al. (eds.), *Integrated Science*, DOI 10.1007/978-0-387-84853-2_3,
© Springer Science+Business Media, LLC 2009

movements, science and technology, education, social theory and cultural studies, Professor Aronowitz is Director of CUNY's Center for the Study of Culture, Technology and Work at the Graduate Center.

Dr. Duderstadt has emphasized the need to preserve the core mission of the university while accepting the realities of technology transfer and the demands of industry. In general terms, how important is it to keep separate the goals of higher education and relevant business interests?

The current debate over the reform of higher education appears indifferent both to the historic function of U.S. universities and the broader ideological, economic, and political issues that have shaped them. Against the encroaching demands of a market-driven logic, a number of progressive educators have argued forcefully that higher education should be defended as both a public good and an autonomous sphere for the development of a critical and productive democratic citizenry.[1] Higher education represents for many a central site for keeping alive the tension between market values and those values representative of civil society that cannot be measured in narrow commercial terms but are crucial to a substantive democracy. Education must not be confused with training, suggesting that educators resist allowing commercial values to shape the purpose and mission of higher education. Richard Hoftstadter recognized the threat that corporate values pose to education, arguing that the best reason for supporting institutions of higher education "lies not in the services they can perform . . .but in the values they represent."[2] The values of justice, freedom, equality, and the rights of citizens as equal and free human beings are central to higher education's role in educating students for the demands of leadership, social citizenship, and democratic public life.[3]

What factors have shaped the changes in the ways that universities and industries position themselves?

The ascendancy of corporate culture in all facets of U.S. life has tended to uproot the legacy of democratic concerns and rights that historically defined the stated

---

[1] See S. Aronowitz and H. A. Giroux, *Education Still Under Siege* (Conn.: Westport, 1992); R. Martin (ed.) *Chalk Lines: The Politics of Work in the Managed University* (Durham: Duke University Press, 1998); S. Aronowitz, *The Knowledge Factory: Dismantling The Corporate University And Creating True Higher Learning* (Boston: Beacon Press, 2000); and H.A. Giroux, *Impure Acts: The Practical Politics of Cultural Studies* (New York: Routledge, 2000).

[2] R. Hofstadter and C. D. Hardy, *The Development And Scope of Higher Education in the United States* (New York: Columbia University Press, 1952), 134.

[3] See: R. Hofstadter, *Anti-Intellectualism In American Life* (New York: Vintage, R. 1963); R. Hofstadter and C. D. Hardy, *The Development And Scope of Higher Education in the United States.* (New York: Columbia University Press, 1952); R. Hofstadter and W. P. Metzger, *The Development of Academic Freedom in the United States* (New York: Columbia University Press, 1955); and E. Press and J. Washburn, "The Kept University." *The Atlantic Monthly* (2000) 285(3): 39–54.

mission of higher education.[4] Moreover, the growing influence of corporate culture on university life in the United States has largely undermined the distinction between higher education and business that educators such as Hoftstadter wanted to preserve. As universities become increasingly strapped for money, corporations are more than willing to provide the needed resources, but the costs are troubling and come with strings attached. Corporations increasingly dictate the very research they sponsor. At the University of California at Berkeley, for example, business representatives are actually appointed to sit on faculty committees that determine how research funds are spent and allocated.

Equally disturbing is the emergence of many academics who hold stocks or gain other financial incentives in the very companies sponsoring their research. As the boundaries between public values and commercial interests become blurred, many academics appear less as disinterested truth seekers than as operatives for multinational interests. Yet there is more at stake than academics selling out to the highest corporate bidder. In some cases, academic research is compromised; corporations routinely censor research results at odds with their commercial interests.[5]

Are you suggesting a conflict of interest exists between the academic practices of faculty and the corporations who sponsor their research?

As a 1996 study published in the Annals of Internal Medicine found, 98 percent of the papers based on industry-sponsored research reflected favorably on the drugs being examined, compared with 79 percent of papers based on non-industry-funded research.[6] Press and Washburn have also provided examples of companies that have censored corporate-sponsored research papers by removing passages that highlighted unfavorable results or negative outcomes.[7] It gets worse. As large amounts of corporate capital flow into universities, those areas of study that do not translate into substantial profits get marginalized, underfunded, or eliminated. Hence, we are witnessing both a downsizing in the humanities as well as the increasing refusal on the part of universities to fund research in public health or science fields that place a high priority on public service. The new corporate university appears to be indifferent to ideas, forms of learning, and modes of research that lack commercial value.

---

[4] S. Slaughter and L. L. Leslie, *Academic Capitalism: Politics, Policies, and The Entrepreneurial University* (Baltimore: Johns Hopkins University Press, 1997).

[5] M.K. Cho, "Secrecy and Financial Conflicts in University-Industry Research Must Get Closer Scrutiny." *Chronicle of Higher Education* (1997)43(7): B4–B5.

[6] M.K. Cho and L.A. Bero, "The Quality of Drug Studies." *Annals Of Internal Medicine* (1996) 124(5): 485–89.

[7] E. Press and J. Washburn, "The Kept University." *The Atlantic Monthly* (2000) 285(3): 39–54.

Does a changing economic climate really affect the academic priorities of universities and those of their students?

Within the neo-liberal era of deregulation and the triumph of the market, many students and their families no longer believe that higher education is about higher learning. Instead, the university offers a means of gaining a better foothold in the job market. Colleges and universities are perceived – and perceive themselves – as training grounds for corporate berths. Corporate culture has also re-formulated social issues as largely individual or economic considerations, canceling out democratic impulses by either devaluing them or absorbing such impulses within the imperatives of the marketplace. As our civil institutions' power is reduced in ability to make corporate power accountable, it becomes more difficult within the logic of the bottom line to address pressing social and ethical issues.[8] This shift suggests a dangerous turn in U.S. society that threatens both our understanding of democracy as fundamental to our basic rights and freedoms and ways to rethink and re-appropriate higher education's meaning and purpose.

In the name of efficiency, educational consultants across the nation advise their clients to act like corporations, selling products and seeking "market niches" to save themselves. Within the corporatist regimes many schools have become, management models of decision-making are replacing faculty governance. Once constrained by the concept of "shared" governance, administrators within the past decade have taken more power and reduced faculty-controlled governance institutions to advisory status. Given the narrow nature of corporate concerns, it is not surprising that, when matters of accountability become part of the language of school reform, they are divorced from broader considerations of social responsibility. As corporate culture and values shape university life, corporate planning replaces social planning, management becomes a substitute for leadership, and the private domain of individual achievement replaces the discourse of public politics and social responsibility.

Missing from much of the corporate discourse on schooling are analyses of how power works in shaping knowledge, how teaching broader social values provides safeguards against turning citizen skills into training skills for the workplace, or how schooling can help students reconcile the seemingly opposed needs of freedom and community to forge a new conception of democratic public life. In the corporate model, knowledge becomes capital, a form of investment in the economy, but appears to have little value when linked to the power of self-definition or the capacities of individuals to expand and deepen the scope of freedom and democratic identities, rights, and social relations. Nor does such a language provide the pedagogical conditions for students to engage knowledge critically as deeply implicated in issues and struggles concerning the production of identities, culture, power, and history. It offers no way of recognizing that education must be more than simply a form of training, because it always pre-supposes an introduction to and preparation

---

[8] Z. Bauman, *In Search of Politics* (Stanford, Calif.: Stanford University Press, 1999).

for particular forms of social life, a particular rendering of what community is, and what the future might hold.

Given the competing interests between academia and industry, how would you then describe the mission of today's university?

The current debate about the university's mission is centered crucially on the curriculum. Since 1979, when Harvard's administration re-imposed an attenuated core curriculum, most colleges and universities have enacted a mélange of similar reforms such as distribution requirements, "computer" literacy, and quantitative reasoning. Higher education is in the throes of a second stage in curriculum "reform." Leaders in this movement ask about the purpose and goals of education. Feminists and educators of diverse racial backgrounds responded to the imposition of core curricula that resuscitates the traditional literary canon as a site of privileged learning by insisting on the inclusion of global, post-colonial, and otherwise marginalized literatures and philosophy. Yet the so-called multicultural or diversity curriculum only peripherally addresses the central problem that afflicts private as well as public universities. Executive authorities in and out of the institution have commanded that schools justify their existence by proving value to the larger society. In most cases, they have meant business interests. In turn, consistent with one of the powerful strains in the history of higher education to subordinate higher learning to practical interests, presidents and provosts are inclined to seek a "mission" that translates as vocationalization, which entails leasing or selling huge portions of its curriculum and its research products directly to companies.

As a result, public and private research universities are dusting off a suggestion from Clark Kerr, former chancellor of the University of California at Berkeley, that undergraduates as well as graduate students should be recruited to participate in the research activities of the professoriate, especially in the sciences.[9] Like sports and many other activities, research demands a considerable time commitment from the practitioner. Universities in the California system – including San Diego and Irvine – are reducing the course requirements of science and technology majors in the humanities and social sciences so they can more accurately mimic the practices of the great private technical universities. This shift, of course, raises the question of whether public universities, as public goods, should maintain their obligation to educate students in citizenship as well as job skills.

Has this shifting focus onto industry-relevant research had a discernable effect on the curriculum?

In the late 1970s, during legislative hearings, the chair of the higher education committee of the California State Assembly and other committee members expressed concern that faculty members were avoiding undergraduate teaching in the service of their research and that state universities were slighting programs aimed at

---

[9] C. Kerr, *The Uses of The University* (Cambridge, Mass.: Harvard University Press, 1963).

educating for citizenship. University administrators appeared to bow to legislators' stern warning that if they did not alter the situation their budgets would feel the heat. As with all attempts by legislatures to micro-manage education, it did not take long for the administration and the faculty to regain lost ground. Today, most public campuses in California are monuments to technoscience. With few exceptions, mostly at the undergraduate level, the humanities and social sciences are gradually being relegated to ornaments and service departments.

In our smaller, less prestigious schools, the forms of privatization and vocational-ization are far more explicit. So, for example, Bell Atlantic has developed relation-ships with public community and senior colleges throughout New York on condition that the schools agree to enroll and train students for specific occupations needed by the company. In most cases, no money changes hands, but the schools benefit by additional enrolment and other gains. For instance, this practice allows state universities an easy means to demonstrate to the legislature and other politicians that they play a role in increasing worker productivity and enhancing economic growth. On this basis, they argue, legislators should reward them with additional funds. Furthermore, the company gains because its employees learn occupational skills that often lead to upgrading, and the company is able to transfer the costs of training to the public.

The question at issue is whether schools should forge direct corporate partner-ships and, in effect, sell their teaching staff and curriculum to vocational ends. In the occupational programs we have examined, the liberal arts, especially English and history, play a service role; at Nassau Community College in Long Island, stu-dents are required to take a course in labor history and their English requirement is confined to composition. Otherwise, the remainder of the two-year curriculum is devoted to technical subjects of direct applicability to the telephone industry. Put more broadly, third-tier public colleges and universities are under pressure to reduce their humanities and social sciences offerings to introductory and service courses to the technical and scientific curriculum. In effect, the prospective English or soci-ology major faces a huge obstacle to obtain a degree in their chosen discipline, because there are often not enough electives to fulfill the major's requirements. As a result, we can observe the rush to mergers of social sciences departments in many third-tier public schools.

How has this emphasis on a practical curriculum affected the liberal arts?

This consolidation has not, in most places, resulted in resurgence of the liberal arts. For example, at Cameron State University in Lawton, Oklahoma, the two philoso-phers on campus are now in the social science department. They teach applied courses such as business ethics. To maintain viability, the department offers a major in occupational specializations of social welfare, a vocational sequence designed to train counselors, and a program to produce low-level professionals in the criminal justice system, a thriving industry in the state. Lacking a social and a political theorist on the faculty, these required courses are taught by a criminologist. With almost 500 majors, the 11 full-time members of the department each teach more than 120 students in four course loads a semester. They also serve as academic and professional advisors for bachelor's and master's students. Many courses are taught by adjunct faculty members. Because the university has many business majors, a favorite of dozens of these small schools, the humanities and social science departments are crucial for fulfilling the shriveling "breadth" requirements.

Economic pressures as much as ideological assaults on the liberal arts account for the curricular change in process in public higher education. The student and his or her family feel more acutely the urgency of the race for survival. The relative luxury of the liberal arts might be reserved for the few who are liberated from paid work during their college years. The consequence is that the human sciences are squeezed from the bottom as well as the top as students demand "relevance" in the curriculum and lose their thirst for reflection.

Are there any examples you can cite that stand in opposition to this convergence between the liberal arts and the objectives of industry?

It may be safely declared that only in the larger cities – and then not uniformly – have faculty members and students successfully defended the liberal arts. At City University of New York (CUNY), a decade of determined faculty resistance has slowed, but not reversed this trend. In part, this minimal success is due to the almost complete lack of media coverage of these issues at CUNY and countless other campuses across the nation.[10] As the new century dawns, CUNY administrators are following the lead of other public and private universities, preparing its version of distance learning. This educational formula is one of the more blatant efforts to end the traditional reliance on classroom learning in favor of a model using technology to produce more standard packages of pre-digested knowledge. It is also an answer to the fiscal crisis suffered by many public schools, because the style of learning reduces the number and proportion of expensive, full-time faculty members. Distance learning instead favors adjunct faculty members. It also transforms brick and mortar into cyberspace, reducing building and maintenance costs and, through standardization, eliminating the mediation of a critical intellectual to interpret transmitted knowledge. The latter saving does not refer as much to cost as to the centralization of political and social control.

Can you provide any specific examples of curricular impacts?

Neither the discourse nor the practices of critical learning are abroad in public higher education, except as the rear-guard protests of an exhausted faculty and a fragment of the largely demobilized student body. Furthermore, as recent changes at the University of Chicago attest, leading private schools are under pressure to dilute their offerings. Blind sided by the rebellions of the 1960s, many educators went along with student demands for ending requirements and ended up with the

---

[10] E. Press and J. Washburn, "The Kept University." *The Atlantic Monthly* (2000) 285(3): 39–54.

marketplace in which demand-driven criteria determined curricular choices. In other words, neo-liberalism entered the academy through the back door of student protest. For progressive educators, however, the task remains unchanged. Rather than accept the flaccid "breadth" requirement of many universities that claim to offer a core, these educators must demand a rigorous and coherent core curriculum in which the history and diversity of Western and Eastern knowledge are critically examined. This demand is not a call to revive the Great Books for the top 20 universities, as suggested by conservative Alan Bloom (1987), but to propose that all students, especially in their first two years, study science, literature, and philosophy in global historical contexts regardless of their institution's position in the hierarchy. As a requisite of any post-secondary credential, this demand is today a radical act. Profit making does not define the meaning of democracy, nor should the laws of the market define the essence of higher education. For higher education to capitulate to the "market" – which arguably wants something else because it is in a panic about an uncertain future – means that training replaces education and that surrendering the idea of higher education as a public good is a necessity. A democratic – as opposed to a commodified – education would acknowledge that public institutions, largely paid for by working- and middle-class people, should promote critical thinking, explore the current meaning of citizenship, and relentlessly pursue democratic appropriation of both Western and Subaltern (marginal) traditions with bold skepticism.

Perhaps it is too early to propose that public higher education be thoroughly de-commodified, that all costs be paid by a tax system that must be made progressive again. Perhaps the battle cry that, at least in the first two years, only science, math, philosophy, literature, and history (understood in the context of social theory) be taught and learned, confining academic and vocational specializations to the last two years, is too controversial, even among critics of current trends. If higher education is truly to become a public good in the double meaning of the term – as a de-commodified resource for the people and as an ethically legitimate institution that does not submit to the business imperative – then we must move beyond mere access. We must promote a national debate about what is to be taught and learned if citizenship and critical thought are to remain, even at the level of intention, at the heart of higher learning.

David L. Kirp, Professor, Goldman School of Public Policy, University of California, Berkeley

Dr. Kirp is author of *Shakespeare, Einstein and the Bottom Line: The Marketing of Higher Education* and many other works related to the relationship between the academy and industry.

Dr. Duderstadt has advocated the need to retain the traditional goals of the university while accepting the realities of market forces in terms of such enterprises as technology transfer. Dr. Aronowitz has argued that such a balance may be impossible and that the university should get out of the business of business. How do you see the relation between the goals of the University and the demands of industry?

While the public has been napping, the American university has been busily reinventing itself. In barely a generation, the familiar ethic of scholarship – baldly put, that the central mission of universities is to advance and transmit knowledge – has been largely ousted by the just-in-time, immediate-gratification values of the marketplace. The Age of Money has reshaped the terrain of higher education. . . .Gone, except in the rosy reminiscences of retired university presidents, is any commitment to maintaining a community of scholars, an intellectual city on a hill free to engage critically with the conventional wisdom of the day. The hoary call for a "marketplace of ideas" has turned into a double-entendre, as the language of excellence, borrowed from management gurus, dominates in the higher-education "industry." Trustees, administrators, faculty, students, business, government – everyone involved in higher education is a "stakeholder" in this multibillion-dollar enterprise.

University administrators used to play Robin Hood, redistributing pots of money, much of it overhead generated by federal military contracts, to support liberal arts programs. But in the name of "responsibility-based management," programs that raise money from outside sources get to keep it. As pointed out in *Harvard Magazine*, disciplines tied to money – either because their subject is wealth, or because they are in close proximity to the wealthy, or because they generate wealth for the university through the transfer of their technology to industry – are handsomely rewarded. Superstars are pursued as ardently as (if with somewhat more modest offers than) Ken Griffey Jr. When Columbia University offered Harvard economist Robert Barro $300,000 plus a raft of perks and the authority to rebuild the department, the story appeared on the front page of the *New York Times'* business section. Less famously but more commonly, universities are investing millions to recruit leading scientists. Columbia was able to hold on to a prize-winning chemist only by promising to buy a piece of property and build a new lab to house a multimillion-dollar piece of equipment. Meanwhile, less lucrative academic pursuits are maintained as museums of the outmoded, and their curatorial faculty is paid a comparative pittance. As recently as a quarter-century ago, approximate parity in salaries was the norm, but today a full professor of English earns no more than a starting assistant professor of accounting.

The dining hall, the bookstore, the infirmary, the dormitory, even the library – more and more, every part of the university that can potentially make money (or, like the library, hemorrhage money) is being "outsourced." So too with teaching: Nearly half of all higher-education faculty, twice as many as in 1970, are part-timers. They are literally "adjunct" – marginal – to the enterprise. As well, half of all new full-time faculty are now hired on short-term contracts, with no chance for tenure, marking a deep change in academic culture. Except at top-tier universities, tenure is vanishing, by stealth rather than by decree, as professors' claims that lifetime employment is essential to maintaining academic freedom lose out to the needs of the "knowledge industry" for a flexible academic work force.

Students, who used to be regarded as acolytes and learners, have morphed into consumers who can be "marketized" – that is, pitched. They are customers whose preferences are to be satisfied, not challenged by intellectual heresies. Students are less idealistic, more vocationally oriented, than ever. A quarter of all undergraduates now major in business, up from just 4 percent in 1970, while enrollment in the social sciences and humanities plummets.

Universities woo prospective students by pitching the quality of undergraduate life as if they were selling time shares, promising apartments rather than dorm rooms, high-tech gadgetry and state-of-the-art gyms. Faculty at some schools feel the pressure to keep grade-point averages high to keep the customers happy. The most academically talented students are avidly recruited by universities anxious for a loftier perch in the academic pecking order. Money is a major talking point, as merit-based scholarships are, more and more, replacing financial aid based on need – a reminder that the market, whether for widgets or higher education, is no respecter of equity.

Responsiveness to new student demands, the emergence of for-profit competition from multisite schools like DeVry Institute and the University of Phoenix, as well as from schools like Jones International University, which exist only virtually – taken together, these developments have made "any time, any place" higher education a near reality. This is the market functioning at its best. But at the same time, the gap between elite universities and mass institutions is widening. As scores of new institutions compete for MBA students, the value of a Stanford or Wharton degree rises. Because of their greater scarcity, these "brands" are worth more.

Struggling liberal arts colleges hire firms to make cold calls for freshman recruits and reinvent themselves as vocational schools in order to survive. Second-tier public universities are obliged to make do with less funding. Meanwhile, the richest institutions grow richer, as money begets money. Endowments soar along with the Dow Jones average, and technology transfer deals grow more lucrative as the leaders in the pack secure nearly $100 million a year in fees and royalties. The higher a college's *U.S. News & World Report* ranking, the easier it is to attract good students. Even as financially hard-pressed colleges rely on tuition hikes to pay the bills, Williams College, awash in endowment, announces – noblesse oblige – that it is freezing tuition levels.

It's Canute's folly to wish or will away these developments. There's no returning to Cardinal Newman's classic *Idea of a University*, where "useful knowledge" is a "deal of trash" – no returning, either, to the world envisioned 80 years ago in Thorstein Veblen's *The Higher Learning in America*, where pure research is the only legitimate pursuit of scholars and "vocationalism" is banished from academe. In reality, such a university never existed in the United States, where the "practical arts" have been a central part of higher education's mission since the launching of land-grant universities in the mid-19th century. What is new and troubling is the power that money directly exerts over every aspect of higher education. There is surely a place for the market in academic life, but the market needs to be kept in its place. The critical question is how to draw, and how to maintain, this line.

Some higher-education initiatives, undertaken in the name of market responsiveness, plainly cross this line. Michigan Virtual Automotive College, the offspring of the University of Michigan and Michigan State, is a good example. Its *raison d'être* is to produce courses that satisfy the demands of the Big Three car manufacturers. Perhaps what's good for General Motors is, after all, good for the nation, but it's not necessarily good for higher education. MVAC courses are open only to employees of the sponsoring firm – so much for the "community" in community college. The curriculum can incorporate proprietary information that instructors are forbidden from using in their other courses – so much for academic freedom. Decisions about what courses to offer are made without any sort of faculty vetting – so much for academic autonomy. No questions would be asked if Ford Motor Company University offered such courses, but the auto manufacturers have farmed out their corporate training, complete with corporate secrets, to a public institution.

Are there Presidential leaders who understand how to draw and maintain the line between academic life and the market?

Visionary university presidents – even articulate university presidents – are a nearly extinct breed: There is no time to contemplate the mission of the institution while constantly raising buckets of money. Among higher education's leaders, only a few – Harvard's former president Derek Bok and Princeton's ex-president William Bowen are the best examples – write intelligently about these matters. Clark Kerr's *The Uses of the University*, first published in 1963 and now in its fourth edition, remains the clearest statement of the purposes of a modern university – the multiversity, as Kerr called it, a "city of intellect." Though Kerr's argument is too obeisant to the practical and insufficiently attentive to pure scholarship and teaching, no one has made the contrary case convincingly.

There is a persuasive case for universal liberal education (as laid out powerfully elsewhere, for example by Martha Nussbaum in *Cultivating Humanity*)...Yet the pressures of the marketplace point elsewhere, toward a world where "knowledge production and transmission must now justify itself in terms of its economic value." Unless mainstream America can be convinced that a liberal arts education is something of intrinsic, not simply economic, value, there will be no slowing the drive to turn higher education into just another market sector. It's not entirely farfetched to imagine that, a generation from now, shares of Yale Inc. and the Princeton Corporation will be bought and sold, just like stock in the University of Phoenix or IBM.

Chapter 4
What Are the Institutional Obstacles to the Integration of Academic Science and Industry?

Henry Riggs, Founding President and Trustee, Keck Graduate Institute

Among other academic and industry experiences, Henry Riggs was the Founding and Former President of the Keck Graduate Institute and President and Chief Executive Officer, Icore Industries, Sunnyvale, California.

M.E. Brint et al. (eds.), *Integrated Science*, DOI 10.1007/978-0-387-84853-2_4,
© Springer Science+Business Media, LLC 2009

Given your background, how would you gauge the similarities and differences between the roles of leadership in industry and in the academy?

Having spent some early years as president of a company and more recent years as an academic president, I am convinced the jobs are startlingly similar and growing more alike with time. However, lest my academic colleagues dismiss my observations as those of just another "for-profit type" seeking to impose unworkable methods on the unique environment of higher education, I must emphasize that not all elements of industry models and processes – particularly those of control and management – can be transferred to academe. The reverse is also true. Additionally, despite many similarities between their core responsibilities, I do not suggest that academic presidents be recruited from industry or industry executives be recruited from academe. I do suggest, however, that the hand-wringing about the exceedingly difficult task of academic leadership is overdone.

What has caused the two leadership types to converge?

I propose that academic and industrial institutions are being reshaped by the same four external forces; rapidly changing markets, heightened competition, new technology, and demands for accountability by multiple constituents. Additionally, each sector is responding to these forces by transforming internal management methods and practices. Incidentally, this trend is particularly evident among technology-intensive companies whose success partly depends upon a cadre of highly educated, creative, and independent professionals created by and found in colleges and universities.

Let's begin with external forces of change. Could you develop the four forces you just mentioned?

1. Rapidly Changing Markets Create Market-Driven Strategies

It is an article of faith in industry that markets change rapidly, particularly in present times. Academic institutions are awakening belatedly to realize that they too operate in markets that are rapidly changing. Companies are much more zealous than they once were about "getting close" to customers by understanding and satisfying customers' needs. In many companies, all corporate strategies are built on this base; their capabilities are unambiguously focused on satisfying customers.

Higher education follows in these footsteps. Academic institutions operate in many markets: the labor markets for faculty and staff, the student recruitment market, the philanthropic market. [The role of] the academic leader is the same as that of the industrial marketing manager: Get close to the customer (donor), understand his or her needs (hopes and desires), and match them to the needs (opportunities) of the institution.

2. Heightened Competition

Competition is a way of life for industry. However, the arrival of global markets and global suppliers has made it more intense. In the academic world, heightened competition is a new reality, sprung up among colleges and universities to an extent and in ways that some view as unseemly.

Ask any admissions officer, from the most elite college to the most comprehensive state institution, about competition to attract the best students. Despite differences in the definitions of "best students" among institutions and even constituencies, the competition to attract them is nearly universal. While the general public may not be fully aware of the price competition that has come to higher education (particularly among private institutions), admissions offices are resorting to a blizzard of recruiting techniques: discounted tuition, special scholarships, and other deals. Tuition's sticker price only indicates where the bargaining begins. Competition extends to the research dollar, as research support from federal agencies and industrial research labs shrinks. This is also true for the philanthropic dollar: not only are social agencies and cultural institutions relying more on gift funds as governments reduce funding, but public institutions also are entering the fund-raising business in record numbers. Institutions are competing fiercely to set and achieve very ambitious capital campaign goals – and competing for the talented, responsible, experienced development officers, who are in short supply.

Academic fund-raising techniques bear a startling resemblance to those employed in consumer and industrial marketing: Understand the values important to the customer! Link price to value! Segment the customers based upon geography and special interests, use direct mail and so forth.

Additionally, colleges and universities compete for faculty, although higher education traditionally has felt relatively immune from labor market forces. However, those markets are placing differential values on different skills and abilities, just as is the case in industrial markets. An example of this is the increasing divergence of faculty salaries across disciplines.

3. An Avalanche of New Technology

Leaders in industry and the academy are struggling to understand and apply cost-effectively an avalanche of new technology much of it to operations and marketing. The technology emphasis in industry has been on improving productivity, but leaders increasingly realize that technology yields an equal payback in improved quality. For faculty, the motivation to adopt new technology has been quality improvement-more effective teaching and learning. Only as financial pressures intensify in higher education do academic leaders view technology as a possible source of reduced costs. The phrase "teaching productivity" is still anathema to some faculty members, but it inevitably will become less so.

A key challenge for leaders in both arenas is to realize – and then convince their organizations – that productivity and quality improvements are not competing or contradictory priorities. Not only are they compatible, but they are mutually reinforcing.

4. Demand for Accountability from Multiple Constituencies

The world has long empathized with college and university leaders in their struggles to cater to multiple constituencies: students, parents, faculty, alumni, trustees, staff, donors, lenders, research funders, government regulators, the appropriations committees of state legislatures, and many others. Indeed, the need to orchestrate the often conflicting demands of these many constituencies usually is cited as a major factor in differentiating the academic president's job from that of his or her industrial counterpart. To be sure, one peculiarly academic constituency is tenured faculty; the president's accountability to his or her faculty presents special challenges.

Nevertheless, in this instance, industry increasingly is moving toward the academy. We now hear that industry must be responsive to multiple stakeholders, including employees, members of the surrounding community, government regulators, suppliers, joint-venture and alliance partners, and customers – in addition to the traditional stakeholders of shareholders and debt-holders.

In both worlds, the demands for accountability become more shrill by the day-accountability that reaches well beyond finances. But as for financial accountability, the scrutiny from trustees and others finally seems to be causing us to scrap – or at least greatly modify – traditional fund accounting, an accounting method that obscures far more than it ever illuminates.

Given that these four external forces are evident in industry and academe, are their operational responses similar?

We may use different vocabulary to describe such management practices, but the organizational assumptions that underlie them and the day-to-day behavior changes they elicit are more alike than not. Nor are the similarities confined to support staff or non-academic activities. Rather they apply to the mainline teaching and research activities as well.

What are some of these similarities?

Empowerment

This word has become a cliché in business. Industry has come to recognize the benefits: delegating authority and responsibility well down into the organization. It has taken business a long time to discard the management structures promulgated by Frederick W. Taylor at the turn of the century and grasp the obvious: The person who performs a task is usually in the best position to figure out how to do it better. Empowerment requires top management to share key tasks with colleagues throughout an organization.

College and university leaders have been doing this with faculty for generations; in the academic world, it's called "shared governance." Academic institutions long have acknowledged that their strength lies in those who do the "line" work of teaching and research. Faculty members are usually the source of the best new ideas, and they have much of the responsibility for implementing new initiatives. That sounds like empowerment to me.

To the extent that a company has highly competent and dedicated employees – from office to scientists and technologists to marketing directors to the rank and file – those employees expect to be involved in key decisions: They expect a real form of shared governance for precisely the same reasons that faculties insist on shared governance. Chief executive officers (biotechnology companies are as unlikely to dictate investigative paths to their companies' research scientists as academic presidents are to outline a faculty member's precise scholarly research or course syllabus).

There is a critical difference, however. Tradition and constitutions *require* academic chief executives to seek advice and counsel of faculty on virtually all key matters, not solely those affecting curriculum, pedagogy, and research. Industrial chief executives retain the *option* to empower and generally do so regarding a more limited set of corporate issues.

Empowerment in the academy extends to students as well. Old traditions of "command and control" teaching-lecturing to students and testing them on the lectures – are giving way to coaching, mentoring, and collaborating to encourage students to take responsibility for their own learning.

Industry has witnessed the enormous increase in the use of teams for solving problems, developing new products, and planning marketing campaigns – among other functions. The academic world is replete with faculty committees, for which it is frequently and appropriately criticized. When a committee is constituted as a task force with specific and consequential goals, a time frame, and members who are given distinct roles, the committee is simply a team by another name. As industry responds to a challenge or problem by setting up a team, so the academic world inevitably forms a faculty committee.

Surely one of the major criticisms of the industry approach to higher education is the focus on the former's profit motive. Would you say that higher education has a similar motive?

[The question is actually one of] imbalanced priorities. Corporate America is criticized frequently and vehemently for overemphasizing earnings per share one fiscal quarter at a time. The argument goes that too many companies sacrifice long-term priorities-research, market development, modernizing facilities – to the expediency of near-term profits. On the other hand, the long term *is* irrelevant unless the company and its managers survive the short term. Balance is key. Every industrial leader knows it, but developing consensus on just what constitutes the proper balance is no simple matter.

The academy is roundly criticized for overemphasizing faculty research at the expense of teaching activities, particularly with undergraduates. When the phrase "scholarly activity" is substituted for research – an appropriate substitution for most faculty members – the conflict between short-term and long-term priorities becomes more apparent: teaching timelines are short compared with research timelines. A faculty member who is today a fine classroom teacher must remain intellectually active and alert to sustain or enhance his or her teaching excellence. As teaching becomes less an activity of lecturing and more one of mentoring or coaching, the greater the importance of a faculty member's dedication to building his or her intellectual capital.

Interestingly, the public reverses its arguments in the two realms: Industry, some pundits suggest, should take a longer view (invest in research and development), and the academy should focus more on the short-term (focus on undergraduate teaching). Balancing the conflicting long-term and short-term priorities is an art to be practicel by both sets of leaders.

What other management challenges do leaders in both sectors confront?

As both industrial companies and universities grow and diversify, they tend to decentralize into entities that become decreasingly similar. In industry, companies evolve from functional organizations to separate divisions to groups. Over time, some corporations take on the characteristics of holding companies.

Colleges and universities evolve in much the same way: from academic departments to separate schools within the university, perhaps to multiple campuses. The individual units become more heterogeneous: the activities of the English department in the school of arts and sciences, for example, are vastly different from those in the cardiology department of the medical school.

Large research and land-grant universities pursue substantially less focused missions than those in the corporate world-giant multi-industry, multi-national concerns. The argument holds, then, that decentralization challenges academic leaders even more than it does their peers in industry. Ask any university president where the power, the authority, and the resources reside on his or her campus; his or her answer will be unequivocal. They are in the schools, particularly the professional schools and the athletic departments-units that raise their own funds and resent sharing them with the central administration as a "tax" for being part of the university. The business school dean's office may be across the street from the president's, but often it may as well be in Tokyo or Karachi, or Rio.

Decentralization creates challenges of coordination and control equally for the academic leader and the industrial executive. They struggle to develop appropriate information and reporting systems; they agonize over the right balance between providing incentives (not just monetary ones) and exercising control; and they labor to reinforce the organization's central mission and culture in far-flung and disparate units. Both industry and the academy pay a price for this decentralization. For example, neither is immune to the occasional ethical lapse, financial surprise, or public-relations debacle.

Constraints on Labor Flexibility

A common, although admittedly less frequent, lament of industry is that labor unions restrict a company's flexibility to change operational methods, reassign employees, alter incentive compensation plans, and augment or reduce the labor force. In some countries (Germany, and to some extent Japan), these constraints – some in the form of governmental regulations – are powerful restrictions on managers' freedom to make decisions regarding the work force.

Unions are not absent in colleges and universities: faculty, staffs, and now graduate students often are organized, even if informally. Further, the academy operates with the additional constraint of faculty tenure. Just as most industrial executives would prefer to operate in a non union environment, most academic leaders would not invent tenure if they were designing a higher education system from scratch. But as industry has learned to accommodate labor unions, so the academy accommodates tenure. The long probationary period of typically six or seven years for faculty and the gravity of the tenure decision (are others prepared to embrace the candidate as a colleague for 30 or 40 years?) are powerful quality-control mechanisms not apparent to critics who focus solely on tenure's lifetime employment guarantee.

Compare this situation with the difficulty that many industry managers have in parting company with a mediocre, barely satisfactory employee who survives year after year. Many such employees achieve de facto tenure particularly in large, stable organizations. They face little risk of dismissal, short of the entire enterprise failing or going through the now commonplace practice of downsizing. Moreover, in this age of widespread litigation, employees in all types of organizations can use a heavy arsenal of legal weapons to help them keep their jobs: alleging age discrimination, unjust dismissal, or gender or ethnic bias, among them.

Still another form of "corporate tenure" is the industrial practice of granting "golden parachutes" to senior executives (top academic administrators sometimes are granted these arrangements as well). One reason this practice has evolved may be loosely analogous to the reason tenure arose in the academy: fear of abrupt and capricious dismissal.

Weakening of the "NIH" Syndrome

A common shortcoming shared by industry and academe is the quick dismissal of all ideas and practices "not invented here (NIH)." However, as Japanese industrial managers demonstrated to their counterparts around the world the benefits of learning from others, this syndrome – a kind of myopia – is on the wane in industry. Innovative companies actively practice "benchmarking" to search out best practices in other organizations and emulate them. Many alliances and joint ventures have developed domestically and across national borders as companies recognize the benefits of "teaming." Often, the key benefit is organizational learning.

Fortunately but belatedly, higher education is following industry's lead. To date, academic benchmarking has focused on administrative rather than teaching processes, but increasing emphasis on assessing educational outcomes is likely to accelerate benchmarking of teaching and learning processes. Colleges and universities are discovering they have a great deal to learn from other academic institutions, and not solely from the so-called academic elite. Some of the most intriguing innovations have flowered in institutions that don't even come close to making everyone's top 10 list of prestigious institutions! And the academy is discovering the benefits of cooperating. Even the wealthiest of our colleges and universities know they must share and borrow library resources. The consortiums in the Connecticut Valley in western Massachusetts and in Claremont, California have learned to share a host of class offerings, specialized student services, and selected administrative functions. Financial pressure is a great antidote to the "NIH" syndrome.

Recruitment Processes

The academic hiring process for faculty and administrators often appears inefficient. Colleges and universities form multi-constituency task forces, search far and wide, and consume months of time and effort to hire a senior (sometimes even a junior) faculty member, dean, vice president, or chief executive. Without defending all aspects of this cumbersome process, industry is adopting some of its more useful aspects. Searches in the industrial world have become broad and lengthy, candidates often are subjected to many interviews by senior and peer members of the organization, and peer evaluations play an important role in final hiring decisions.

Leading Versus Managing

If industry has led the academy in implementing many changes, in operational practices and behaviors, the academy has led the practice of good leadership. All management literature and development programs, whether geared to industry, not-for-profit organizations, government, or other entities, stress that presidents must lead, not just manage. As Warren Bennis says in his 1993 book, *An Invented Life* (pp. 75–78), "In contrast to just 'good managers', true leaders . . .affect the culture, are the social architects of their organizations, and create and maintain values . . ." The standard criteria for choosing top-level managers are technical competence, people skills, conceptual skills, judgment, and character. And yet effective leadership is overwhelmingly the function of only one of these: character. Successful academic presidents have emphasized leadership over management for two reasons. First, as noted earlier, they must respond to – and in a real sense, depend upon – a broad range of constituencies. Second, presidents could not manage most of these constituencies if they wished. Of faculty, students, alumni, donors, legislators, the public, trustees, and administrative staffs, the academic president manages – in a command and control sense – only the last. The rest can and often do thumb their noses at directives. They must be led; inspired, coaxed, encouraged, cajoled, supported, and coached.

Leadership has become more widely recognized as the key responsibility for industrial chief executives as they realize the benefits of (1) being more responsive to many quite unmanageable constituencies, including customers, suppliers, employees, directors, local community members, joint-venture and alliance partners; and (2) empowering employees and delegating responsibility to an unprecedented extent.

Academic executives may have pioneered the art of leading constituencies, but in some ways, industrial chief executives have been quicker to focus their leadership skills on adapting organizational cultures to new competitive, financial, and technological realities. Such leadership has resulted in a faster pace of change, particularly among the most prestigious and revered institutions: top industrial companies have sought to adapt their corporate cultures aggressively, whereas the most elite academic institutions have been among the least adaptive.

Fortunately, the job of president is more interesting, more challenging, and more satisfying when the key task is leading. That condition is as true in industry as it is in the academy. This brief catalog of common management behaviors and methods is not exhaustive. Both academic and industrial organizations are now – or soon may be – involved in downsizing, struggling to turn fixed expenses into variable expenses, worrying about asset management, engaging in image management, public relations, and a variety of ceremonial functions.

So far you have spoken of the similar managerial and institutional challenges confronted by industry and the academy. Are there also persistent differences?

[Yes,] despite the many parallels, certain profound and entrenched differences between the academy and industry must be recognized and respected, especially when we consider adopting industrial models of organization and management. Some of those differences are:

Unclear Customer Set

Unlike successful, profitable companies, higher education's "customer set" often is unclear and shifting. In one view, students are the customers; in another, they are perceived as "raw material" for the process of higher education. Further, the market customer may not be the person who pays for the service, and in the case of institutions that have the luxury of employing selective admissions policies, the supplier may refuse service to many more student-customers than it serves and be unwilling to add capacity. Higher education administrators view faculty and donors, as well as students, as customers. For public institutions, the list expands to include elected representatives.

Quality Assessment

In industry, the chief arbiters of quality clearly are customers; in higher education, faculty are the chief arbiters of quality. However, as noted earlier, higher education may be entering a new era in which leaders pay greater attention to the feedback about quality they receive from various constituencies.

Measurement of Outcomes

In higher education, we still are struggling to assess educational outcomes – quality of the product – in meaningful ways. Should we use standardized tests to measure learning? Are job placements a reasonable outcome measure, given that most institutions educate students for life rather than for specific jobs? How can we qualitatively measure the more subtle purposes of higher education – for example, developing responsible citizens and acquainting students with their own and other cultures?

Nevertheless, our publics demand better assessment and understandably so. Our traditional measures of quality focus largely on inputs: number of faculty, class size, volumes in the library, and so on. In the absence of reliable and clear outcome measures, colleges and universities tend to organize and conduct their affairs with an eye to satisfying faculty; without clearly defined objectives, faculty adopt objectives that are to their benefit.

Industry has far fewer problems in measuring its outcomes. While the "bottom line" may put too much emphasis on short-term results, the "bottom line over time" is industry's fundamental and unmistakable measure of success.

Accreditation

Peer review is the essence of higher education accreditation. Accreditation, which is nominally voluntary, is deemed necessary because of higher education's public-service responsibilities. While regulation of industry may be roughly parallel to accreditation, and virtually all professions are licensed by panels consisting of peers, academic accreditation is far more pervasive – and unfortunately often more political.

Academic Freedom

The principle of academic freedom is strongly protected by the culture of higher education and the legal constitutions of colleges and universities. Academic freedom has no counterpart in industry. One of the byproducts of academic freedom is frequent, open, and vociferous (occasionally uncivil) criticism of management, particularly by faculty and students. Of course, similar contentiousness can arise in industry when labor unions are strong and labor-management communications break down. An interesting twist: arguably, the exercise of academic freedom often breeds political correctness, which itself compromises the freedom of expression it seeks to protect.

Resistance to Change

While resistance to change is hardly absent from industrial organizations, it is far more rampant on college and university campuses, in part because the prevailing academic culture requires decision making by consensus. Two of the academy's constituent groups – students and alumni – reinforce conservatism, but the faculty and many administrators share it as well. Their resistance is understandable: they have not needed to change. They sought scholarly careers in part because these careers were secure, stable, predictable, and involved few strict deadlines. Administrators' natural conservatism is sustained further by two factors: (1) income stability in most elite private institutions (at least for the present) and until recently in most public institutions and (2) the resulting absence of financial pressure for radical transformation.

Resistance to change also is reinforced by the important role that tradition plays in the life of an academic institution. Academic presidents see themselves as part of their institutions' traditions, and their plans for the future must respect the past and assure institutional longevity. Some critics argue that academic leaders are unduly constrained by their institutions' traditions. By contrast, an industrial chief executive who is less bound by tradition may see many positive benefits arising from a sharp break from past practices, and may view the future merger or acquisition of the company as a logical and desirable outcome.

However, a parallel exists between the academy and at least one notable segment of industry-public utilities. With deregulation, utility companies are being plunged into turmoil as they try to adapt to a competitive environment. Many managers are working hard to change their organizational cultures. As higher education prepares for the changes of the next decade or two, administrators may discern useful lessons from their utility-industry colleagues.

Ironically, academic institutions have not become learning organizations devoted to continuous change and improvement, despite the opposite expectations they have for their students.

Fragmentation

With great diversity across institutions and within segments of the academic market (liberal arts colleges, engineering programs, doctoral programs, and the like), no single institution has a market share that approaches domination even of a single market-segment, nor, one might argue, does a single institution dominate a national and international market. Moreover, institutions typically are not motivated to grow or to gain market share.

This fragmentation partly accounts for the dramatic reshaping of higher education for the last several decades, even as individual colleges and universities were changing slowly and incrementally. The sum of these incremental changes of growth in total enrolment, an increasing emphasis on adult students, executive programs, satellite campuses or remote delivery, and a shift toward career-oriented majors, is greater diversity of missions among the more than 3,000 academic institutions in this country than was the case in the first half of this century.

No Relevant Capital Market

In industry, mergers, takeovers, and leveraged buyouts redistribute capital among winners and losers, and the ever-present threat of these actions encourages managers to take corrective action before a fiscal crisis ensues. In the academy, no similar capital market operates (although the occasional merger occurs), and the well-endowed institutions are particularly well shielded from fiscal threats.

Further, because industrial concerns have access to well-organized capital markets for equity and debt, they can alter their capital structures to implement a revised strategy more rapidly than can their academic counterparts.

What lessons would you draw from your analysis of the similarities and differences between the academy and industry?

The widespread, erroneous view that industry and the academy are worlds apart, subject to very different external forces that require sharply different management methods and behaviors, deserves correction. When business executives assume roles as higher education trustees, and when senior academic administrators accept membership on corporate boards, they need not and should not assume their career experiences are irrelevant to their new roles.

Senior academic administrators would do well to devote more time to understanding management and leadership practices in industry and considering how they apply, perhaps with modification to their academic institutions. Some critics of higher education are prone to paraphrase Henry Higgins in My Fair Lady in decrying, "Why can't universities and colleges be run more like businesses?" My response: they can and are. We need to remain sensitive to the distinct differences – in mission, constituencies, and measurement of outcomes – between the academy and industry. But the similarities between the academy and industry, and thus the lessons available from across the gulf that has separated them for too long, deserve at least as much attention.

Finally, let me suggest a sobering observation for leaders in both camps. I can think of no revolutionary new product or service – one that spawns a new industry – that has emanated from an organization ensconced in a related market or even from that market's dominant suppliers. Simply put, invention, innovation, and substantive change come from life outside. A plethora of examples proves my point: mini-computers, personal software, overnight package delivery, airport hotels, microwave ovens, and distance learning are only a handful. Notable examples in the academic world are two for-profit academic enterprises; DeVry Institute and the University of Phoenix.

Schumpeter's creative destruction principle often driven by technology, a key external force noted earlier in this paper, seems to be alive and well. As new opportunities emerge in industry and in higher education (teaching in fundamentally different ways to different students at different stages in their lives, for example), will major institutions prove to be more adaptable in the future than they have been in the past?

Leaders in higher education and industry have ignored each other too long, certain that their different environments demanded very different management behaviors. Each camp has tended to drift into a certain self-satisfaction, even smugness, in this view. The time to recognize that these differences have diminished is overdue.

William A. Haseltine, President, William A. Haseltine Foundation

Dr. Haseltine is Chairman of the Board of Directors and Chief Executive Officer of Human Genome Sciences, Inc. He holds a doctorate from Harvard University in Biophysics and was a Professor at Dana-Farber Cancer Institute, Harvard Medical School and Harvard School of Public Health from 1976 to 1993, before joining Human Genome Sciences. He is Editor-in-Chief of the Journal of AIDS, founding editor of the Journal of Regenerative Medicine and is on the Editorial Boards of many other scientific journals. Dr. Haseltine has over 250 publications in the scientific literature. He has been awarded more than 50 patents for his discoveries.

Dr. Haseltine, as the CEO of a Biotechnology company, what kind of training are you looking for in scientists? In your view, do you believe that academic science should emphasize specific disciplines or should they train scientists in an interdisciplinary way?

I come at these kinds of problems from a different perspective. When you are a mature scientist or becoming a mature scientist, the most import attribute is to identify the problem that needs to be solved and then apply whatever tools are needed. That is not a very easy thing to do. First, the hardest thing is to identify the problems that need to be solved. Secondly, how do you train people to use these tools? I don't look at science as compartmentalized. I look at it as a wonderful tool set.

Now I happen to be educated as a chemist, physicist, biophysicist, molecular biologist, virologist, cell biologist. I deliberately picked up as many tools as I could from the very best people I could find with the knowledge that I did not want to be limited by my tools. I think that is a very important concept to instill in a student body. That they should never be limited by tools. For in a sense, that is what these disciplines are. You should try to become as expert in as many of these disciplines as you can. And it is possible for good students to master chemistry, physics, some mathematics today. And they have to know what those foundations are.

Asking you to gaze a bit into your crystal ball, what do you think are the most promising areas of integrated science for the future?

One of the most exciting areas of the future is the intersection of biology and material science. If you look at what has happened over the past 20 years in the field of modern biology – the application of relatively simple robotics and computer technology to biology – it has been transformational. If you are a computer scientist or robotics or materials engineer, you don't think that's a big deal. It is kind of a trivial application to do – it is kind of a klutzy old one. But it has transformed biology. And it has made new industries possible. That's what founded them – a fusion of some of these klutzy techniques of robots moving things around. But that is not the future. The future is where the substance of products are these new materials. The substance of what we sell will be fusion products. And for the material scientist, he will be informed more than ever before, by the example of biology. We are architected at the atomic level. What allows us to function is that our atoms are in a particular three-dimensional array to about a 0.10 angstrom specification. That's a wonderful specification because we know it can work because we seem to work.

People are beginning to think about how to make our material world on the same scale: to use the principles of biology. The products for our material world and the products of our living world are fusing. It is institutions of higher education that have the opportunity to create these new disciplines where you train people to understand what these fusions are. And there is no real barrier between those people who think about structural biology, biological function, and the function of materials, how they assemble – how they self-assemble, how they repair themselves. We are moving from a time when bioengineering provided tools to do our work to a time when it supplies the substance of what we do.

Part III
The Implications of Interdisciplinary Science on Education and Training

Chapter 5
What Are the Long Term Implications of Integration/Interdisciplinary Science on Traditional Disciplines and Their Professional Associations (Turf Wars)?

Steven Brint, Professor of Sociology and Associate Dean, College of Humanities, Arts and Social Sciences at the University of California, Riverside

Dr. Brint's books include *The Diverted Dream* (with Jerome Karabel) (New York and London: Oxford University Press, 1989), *In an Age of Experts* (Princeton: Princeton University Press, 1994), *Schools and Societies* (Thousand Oaks, CA: Pine Forge Press, 1998), and *The Future of the City of Intellect* (Palo Alto: Stanford University Press, 2002).

M.E. Brint et al. (eds.), *Integrated Science*, DOI 10.1007/978-0-387-84853-2_5,
© Springer Science+Business Media, LLC 2009

We have been discussing the changes in the University as a result of increased market forces. From your research (which included analysis of 69 planning documents and interviews with 144 provosts and vice presidents of research of 89 American universities), just how profoundly have institutions of higher education changed?

It is widely acknowledged that American research universities are experiencing a period of profound change. Many see the universities becoming more attuned to market forces, re-engineering themselves to become more efficient, while investing more in areas that attract earnings-conscious students and private research support. As Engell and Dangerfield have put it, "In the Age of Money, the royal road to success (for academic disciplines) is to offer at least one of the following: a promise of money...a knowledge of money...(or) a source of money".[1] The market-consciousness of the "new university" has been celebrated by Liberals,[2] and excoriated by neo-Veblenians,[3] but it has rarely been denied. Key works during the last decade include such titles as *Academic Capitalism, Creating Entrepreneurial Universities, The Enterprise University, Universities in the Marketplace*, and, pointedly, [David Kirp's] *Shakespeare, Einstein, and the Bottom Line.*[4]

However, this emphasis on markets and marketing may be overstated. While acknowledging (and often embracing) intense competitive pressures to hire top faculty and to move up in the rankings, administrators rarely favor the term "entrepreneurial" to describe their institutions. They rarely consider student demand a major influence on the allocation of staffing, and say they hope to retain quality both in disciplines that "promise money" and in those that do not. Indeed, most show less concern with markets than with status. American universities – like universities everywhere – try to enroll good students, hire the best faculty, and improve their comparative position.

For this, there is a straightforward explanation. On the whole, the private research universities in the United States are in a strong financial position, and do not feel compelled to respond reflexively to consumer demand. On the contrary, they can

---

[1] James Engell and Anthony Dangerfield, 'The Market-Model University: Humanities in the Age of Money', *Harvard Magazine*, 3 (May–June 1998), 52.

[2] See e.g., Lewis M. Branscomb and James Keller, *Investing in Innovation: Creating a Research and Innovation Policy that Works* (Cambridge, MA: MIT Press, 1998).

[3] See e.g., Stanley Aronowitz, *The Knowledge Factory: Dismantling the Corporate University and Creating True Higher Learning* (Boston: Beacon Press, 2000).

[4] The cited titles are, respectively, Sheila Slaughter and Larry L. Leslie, *Academic Capitalism: Politics, Policies, and the Entrepreneurial University* (Baltimore: Johns Hopkins University Press, 1997); Burton R. Clark, *Creating the Entrepreneurial University: Pathways to Transformation* (London: Oryx Press, 1998); Simon Marginson and Mark Considine, *The Enterprise University: Power, Governance, and Reinvention* (Cambridge: Cambridge University Press, 2000); Derek Bok, *Universities in the Marketplace: The Commercialization of Higher Education* (Princeton: Princeton University Press, 2003); and David L. Kirp, *Shakespeare, Einstein, and the Bottom Line: The Marketing of Higher Education* (Cambridge, MA: Harvard University Press, 2003).

choose to back programs they like. To compete with private universities, the various State-aided (or "public") universities must follow suit to the extent possible.

The leading research universities have never been as wealthy as they are today. If universities were included among the Fortune 500 companies in 2000, as measured by operating budget, six would have made the list (Harvard, Stanford, Yale, Duke, MIT, and the University of Michigan).[5] Even as State appropriations have leveled or dropped, revenues from tuition, research grants, licensing of technologies, endowment, and annual giving have all increased, often at an extraordinary rate. Universities are no longer as dependent upon tuition, which today accounts on average for less than 25 percent of the typical operating budget.[6]

As their sources of funds have diversified, universities have grown in autonomy and ambition. States and donors are rewarding universities that aspire not to be responsive, but to lead. The position taken by Duke University is characteristic:

> (Private research universities) are resource-intensive places, typically combining large endowments, strong philanthropic support, and external research funding with high tuition. These resources are powerfully (cumulative) in supporting the teaching and research missions of these institutions and their commitment to national and international leadership...*Our overriding goal therefore is to be among the small number of institutions that define what is the best in American higher education.* Certainly Duke can learn from other institutions, but we must also set our own sights and help set the standards for others. This is what leadership means.[7]

Some universities have indicated a preference for pursuing leadership through established professional channels – that is, by improving their standing within the disciplines. Others have indicated a strong interest in following "new directions", less attuned to disciplinary rankings than to making "cutting edge" contributions to new technologies, forms of expression, and social relations. A few universities can do both simultaneously, but this requires resources beyond the reach of all but the top wealthiest. For universities below this level, the more significant choice is not between the Ivory Tower and the marketplace, but rather between building strength in the traditional disciplines and creating new foci of "interdisciplinary creativity".

---

[5] Sources: Steven Brint, Charles S. Levy, Mark Riddle, and Lori Turk-Bicakci, *The Institutional Data Archive on American Higher Education, 1971–2000* (Riverside, CA: University of California, Riverside, 2003); Editors of Fortune Magazine, 'The Fortune 500 Largest American Corporations', *Fortune Magazine*, 143 (8), F-1+. Based on annual operating budgets.

[6] This conclusion is based on an analysis of 16 public and 15 private research universities, which provided full financial reports in the Federal government's Integrated Post-Secondary Education Data System (IPEDS) in 1995. This is the last year for which comparisons between public and private sector institutions are possible, owing to changes in accounting standards for private institutions implemented in 1996. The figure is true both when revenues from auxiliary enterprises are included and when they are excluded. A separate analysis of financial statements from 12 additional public and private universities confirms this figure.

[7] Duke University, Building on Excellence: The University Plan (Durham: Duke University, Office of the Provost, 2001), 9–10 (Emphasis in original).

Among those focusing on interdisciplinary approaches, what have you learned about whether these "new directions" are here to stay? What are some of the reasons proffered for thinking they are permanent and what are some of the reasons for thinking they are mere fleeting phenomena?

[On the one hand,] there are several reasons to think that the "new directions" may turn out to be a passing fashion, rather than a continuing feature. Among the most important are the following:

1. Current visions of technology (and not necessarily basic science) as "the endless frontier" may be the result of what amounts to a historical anomaly, produced largely by the successes of information technology and biotechnology. Some of the many technologies that are fueling the current enthusiasm for "new directions", such as the demand for new homeland security, may not always be as significant as they are today.
2. The receptivity of the Federal and State governments to large scale technology projects could easily wane. Ideological considerations have often limited these efforts in the past, and the current emphasis on strengthening systems of innovation could give way to suspicions that "socialistic efforts" are being used to "pick winners". Budgetary cutbacks could have a similar impact. Indeed, in the wake of recent budget shortfalls, several States, including Alaska, Michigan, New Jersey, and Texas, have reduced or eliminated support for collaboration between universities, governments, and corporations in new technology development.[8]
3. An environment less hospitable to the accumulation of private wealth by the rich could limit the size of future gifts to universities, so effectively limiting the ambitions of university administrators. In this respect, our current era resembles none better than the Gilded Age, before the passage of the progressive Federal income tax in 1911. That age, too, produced both private wealth and privately supported public institution building on a previously unheard-of scale.

Balanced against these possibilities, several arguments support the expectation that the "new directions" will remain important and permanent:

1. Japan has developed and China is developing successful national innovation systems. The European Union has announced analogous plans.[9] International competition will oblige governments and corporations to mobilize intellectual capital to compete in emerging markets.
2. Broad (and less quantifiable) trends in the culture of advanced societies seem to be allied with the "new directions." The work environment now routinely requires multiple skills (political and social, as well as technical) and the capacity

---

[8] Roger L. Geiger and Creso Sa', 'Beyond Technology Transfer: New State Policies for Economic Development for U.S. Universities', Paper presented at the 16th annual conference of the Consortium for Higher Education Research, Porto, Portugal (September, 2003) reprinted in *Minerva*, Volume 43, Number 1 (March, 2005): 1–21, see p. 1.

[9] Ibid., 10–12.

to juggle many tasks at once. The environment of consumption is filled with variety and choice. Information about any subject of interest is at one's fingertips. This daily experience contrasts sharply with the rigid procedures, access-controlled filing systems, hierarchies of experts, narrow job definitions, and highly defined social roles of the bureaucratic organizations that were universally familiar to previous generations of graduates. With the rise of bureaucracy, Max Weber argued, the cultivation of the "generalist type of man", as the highest ideal of the educational system, was displaced by the production of the "specialist type of man". In the new environment, it is worth asking whether the production of the "specialist type" will not in its turn give way to a new "creative type"[10] If so, the institutional changes discussed ... will gain further support from a new representation of the educated self.

---

[10] Max Weber, 'Bureaucracy', in Hans H. Gerth and C. Wright Mills (eds.), From Max Weber: *Essays in Sociology* (New York: Oxford University Press, 1946), 243.

Paul Grobstein, Eleanor A. Bliss Professor of Biology, Bryn Mawr College

Dr. Grobstein is the Director of the Center for Science in Society at Bryn Mawr College, Pennsylvania

Dr. Grobstein, how would you characterize the struggle between emerging and traditional disciplines?

In the middle of the 20th century, the British scientist/novelist C.P. Snow expressed deep concern about what he saw as a split between "two cultures:"

> Literary intellectuals at one pole, at the other scientists . . .Between the two a gulf of incomprehension – sometimes (particularly among the young) hostility and dislike, but most of all lack of understanding . . .This polarisation is sheer loss to us all. To us as people, and to our society. It is at the same time practical and intellectual and creative loss.[11]

Some kind of split persists. As the 20th century drew to a close, the American biologist E.O. Wilson wrote about his hope for the future "if the natural sciences can be successfully united with the social sciences and the humanities."[12]

---

[11] C. P. Snow, *The Two Cultures*, (Cambridge: Cambridge University Press, 1993), 4.

[12] E. O. Wilson, *Consilience: The Unity of Knowledge* (New York: Knopf, 1998), 294.

My own experiences as a scientist, science educator, and parent resonate closely with both Snow's concern and Wilson's dream. But I would characterize the split more generally. As we enter the 21st century, a gap and tension costly to both continues to exist between those who are engaged/comfortable with science and those who are not, both inside and outside academia. My feeling is that this split is illustrative of significant ethical and moral ambiguities in "science" as it is generally understood. As scientists we have a compelling responsibility to see these ambiguities clarified, for our own sake as well as the sake of the human culture of which we are a part. If I had to pick one problem as the most important target for science (and science educators) to address in the 21st century, it would be this one: to clarify the best and most fundamental aspects of science so as to make science a comfortable and accepted part of the shared common story of all human beings. To achieve this, we need to find ways to reduce the perception of science (by both those engaged with it and those not) as a specialized and isolated activity of the few, and create the kinds of bridges that will more effectively link science not only to other academic disciplines but to the non-academic world as well.

Beyond making clearer the actual nature of science, [I see] two additional challenges to completing the task of bridging the "two cultures" gap.

One of these challenges is a matter that depends largely on changes within the existing "scientific community." Indeed, it requires a willingness to change the very definition of "scientific community." Scientists, like all human beings, have a tendency to "tribalism" – an inclination to share observations and stories only with people who are in some sense "like themselves." The main problem with tribalism isn't so much whether members of a tribe are willing to make their observations and stories available to people outside the tribe (which they frequently are) but whether they are also willing to listen to the observations and stories of others, with the potential that those change their own in turn. The scientific community does not have a distinguished record of this latter kind of engagement, and hence it tends by its own tribalism to encourage tribalism in others. If science is actually to become the common property of humanity, scientists themselves are going to need to learn to transcend their own tribal inclinations, to not only entertain the possibility that the observations and stories of people currently outside the community are relevant, but to begin actively valuing them, to genuinely open the "scientific community" to all comers. This will not be easy, but it is in fact very much in line with the core values of science.

Chapter 6
What Are the Implications of Integrated Science for Liberal Arts Education and Pedagogy at the Undergraduate Level?

William Wulf, University Professor, University of Virginia; President Emeritus, National Academy of Engineering

Dr. Wulf was elected President of the National Academy of Engineering (NAE) in April, 1997. Dr. Wulf is a University Professor and the AT&T Professor of Engineering and Applied Science at the University of Virginia. Dr. Wulf is a member of the National Academy of Engineering, a Fellow of the American Academy of Arts and Sciences, and a Corresponding Member of the Academia Espanola De Ingeniera. He is also a Fellow of four professional societies: the ACM, the IEEE, the AAAS, and AWIS.

M.E. Brint et al. (eds.), *Integrated Science*, DOI 10.1007/978-0-387-84853-2_6,   77
© Springer Science+Business Media, LLC 2009

Do you believe traditional approaches to engineering pedagogy are well-suited to tackling integrated science education?

I sense [a] real urgency of engineering education reform. I think we ought to be seeing a watershed change in engineering education – it is not happening. I am very impatient about it and I hope I can communicate . . .why I feel impatient about it. I believe that the way that we will practise engineering and the way that the students we are teaching today will practice engineering are profoundly different from the way that I practiced engineering or my father practised engineering. The problem with trying to describe to you what that change is about is rather like standing too close to a mosaic. I have said, sometimes there are monumental events that kind of cast a sharp knife edge between the way things were and the way things are now. World War II strikes me that way. Before World War II, there was no federal funding of research at universities. After World War II, we built this wonderful mechanism for funding research. The role of women in society dramatically changes across that boundary. In fact, engineering education changes dramatically across that boundary. The notion of the engineering-science model of engineering education comes about because of, frankly, the failure of engineers to contribute as much as scientists did to the war effort.

Are we poised at the brink of such a turning point?

I don't think we are in that kind of a change. I don't see that monumental event. It seems to me that this is much more like the Industrial Revolution. You know, we talk about the Industrial Revolution now as though it was an event. The fact is, it smeared out almost 100 years and it is contemporaneous with a whole bunch of profound changes in society. This is when you get the rise of democracy; this is the rise of rationalism; and there was another great change in university education. The introduction of liberal or secular education comes about exactly the same time. If you were there at the time, you could not have predicted what the world would look like at the end of that time. I think we are in that kind of change.

Can you describe what the changes are that need to be made to engineering education – what content needs to be removed or added?

Curriculum – if you get a bunch of engineers together, there is an oath that we all recite which is because we treat the baccalaureate as the first professional degree. What we must do in the baccalaureate is teach "only the fundamentals." "Only the fundamentals" – you hear that recited over and over again. Well, rubber meets the road when you ask the question what are the fundamentals? And then the mechanicals will tell you something quite different from the civils, and neither one of them will recognize, for example, they sort of agree, because since World War II the fundamentals have included continuous mathematics and physics. That much I think everybody agrees on.

But as I said before, engineering is changing. Information technology is going to be embedded in everything that engineers produce. And discrete mathematics, not continuous mathematics, is the underpinning of information technology. I mentioned biological materials. Biology and chemistry are going to become as fundamental as continuous mathematics and physics. And the fact that engineering is done in this more holistic, team-oriented, multinational global context means that there are a whole set of business and cultural issues which are really fundamental to engineering. You can't practice without them.

Does this mean that "If you want to continue to say that the baccalaureate is the first professional degree, then you have to agree that some of our cherished current fundamentals aren't any more?" Or you have to figure out a way to teach them much more efficiently and effectively.

Ethics has been very important to engineering. Engineers are very much like physicians – first do no harm. We spend a lot of time teaching engineers how to over-design their systems so that they tend to not fail or if they fail, fail safe. How do you cope with the ethics of not knowing what the behavior, what the emerging properties of a system will be? I don't know.

What obstacles exist to implementing the changes to the curriculum that you describe?

Let's talk about faculty rewards. And I don't mean the teaching versus research debate. I happen to be one of those people who believe that most of the time research and teaching complement each other. Most of the people who I know who are good researchers are also good teachers. Good people are good. There are the outliers. But I think we have another problem. Remember I said I believe what engineers do is design under constraint. I happen to think that engineering is an incredibly creative activity. Something we don't advertise very well. In my heart, I believe that engineering is one of the most creative of human activities. If you stipulate that for just a minute, can you think of any other creative activity, on campus, where you don't expect the faculty to practise, to perform that creative activity. The Art Department doesn't promote or tenure anybody who doesn't practice their art. Think about the Music Department. Or even think about the other professions like law and medicine. If you go to medical school, you go on grand rounds with the faculty who is practicing his/her profession. Engineering is the only creative activity that I can think of where, in fact, the faculty is actively discouraged from practicing the profession. And what we wound up with – you know the criteria that we apply for promotion/tenure in universities is essentially derived from the Science Departments. It's research, publication, getting grants, and you'd better teach pretty well too. But, practicing the profession counts for nothing and probably counts against you because it detracts from other things.

So you would suggest that practical experience in industry would actually enhance the ability of engineering faculty to deliver a contemporary engineering education?

I actually had a Dean who would not let one of my faculty take a sabbatical in industry. His image was that there was nothing to be learned from industrial experience and in fact somehow those industrial people were just going to suck out his brains and take out everything he knew. Well, I can tell you, I spent almost 10 years of my life in the private sector and one of the most intellectually challenging things I have ever done in my life was delivering product. It is not just that it is hard, it's intellectually challenging. Going back to the curriculum issue for just a moment, I think one of the things that is really wrong is that we have a curriculum being designed by faculty members who are not practising engineers. I have a great deal of respect for my colleagues at the university. They are wonderful engineering scientists, but very few of them know anything about what the practice of engineering is all about, and so they design a curriculum which is an engineering-science curriculum, not an engineering-practice curriculum.

Is the necessity of engineering education widely shared by people outside of academia?

Before I took this job, I was a Professor at the University of Virginia. As you may know, Virginia was founded by Thomas Jefferson. What you probably don't know, is that Jefferson did not die, he participates actively every day in the decision mechanisms of the university. He was very proud of having founded the university. It was one of three things he put on his tombstone. He didn't mention things like being President of the United States. He founded the university because he believed you could not have a democracy without having an educated citizenry.

Well, I think he would be scared today because we have a citizenry which is not only ignorant of technology, it is proud of the fact that it is ignorant of technology. You know, I go to a cocktail party and someone will ask me what I do and I say I teach computer science and they say, "Oh, I don't understand that computer stuff." Can you imagine asking somebody else what they did and they said they were a Professor of English and you say "Oh, nouns and verbs, I can't . . . ." Engineering schools don't offer technological literacy courses for liberal arts majors. Why not? We could pass on knowledge of not just science and math but the process that takes that knowledge of nature and converts it into the things that profoundly change our quality of life. Think about what somebody in 1899, the average person in 1899, lived like. Think about what an average person in 1999 lived like. All of the differences are engineer products. In 1899, the average life span was 46. In 1999, the average life span was 76. All of that increase is not due to modern medicine. It is almost all due to cleaner water and sanitary sewers – public health. Engineering!!

And yet, "Oh, I don't understand that computer stuff and I am proud of the fact that I don't." Every person who has a liberal education ought to be at some level technologically literate and it's our responsibility to provide the opportunity for that to happen. It is no good to point our finger and say, "You English professors ought to be technologically literate" if there is no mechanism for them to do that.

So you would conclude that engineers of the future will experience a different education than those of the past?

I've tried to indicate . . .that I think the practice of engineering is going to change tremendously and that, therefore, the education of engineers needs to change tremendously. I love this quote, but I don't do it well: Wayne Gretzky, probably the best hockey player that ever lived, talked about the fact that he didn't skate to where the puck was, he skated to where the puck would be. I'm afraid that [current] engineering education is skating to where the puck was.

Donald Kennedy, President Emeritus, Stanford University.

Educated at Harvard University, Dr. Kennedy has spent most of his career at Stanford University as a faculty member, Provost, and President. From 2000–2007, Dr. Kennedy was editor-in-chief of *Science*, the prestigious magazine published by the AAAS.

Do you believe traditional approaches to science pedagogy are well-suited to tackling integrated science education?

I think we have reached a critical stage in the evolution of how we teach science – one that challenges all of the institutions in which we educate undergraduates and has special impact upon the many smaller colleges of high quality that have always trained a disproportionate share of our future scientists. I want to begin the argument by describing the changes that have made this such a turning point for higher education. The first, and most important, has to do with complexity and, concomitantly, cost. We know so much now, and we can do so many things experimentally, that it is very difficult indeed to decide what we should be doing with and for our students. To see what I mean, do the following imaginary experiment. Measure, using weight or number of pages, the most popular biology textbooks used in freshman college and university courses from 1940 to the present. I haven't done it, but I can assure you that there is a monotonic increase now making significant contributions to the problem of backpack-induced postural strain in 18 year olds.

Are changes to integrated science education the result of new knowledge, new methods of teaching existing knowledge, or something else?

There is not only more material, but what one feels one must cover is more difficult, more quantitative and in each domain, more specialized. [As an example] I look at the breadth of coverage in the journal I now edit. Here, from recent issues of Science, are a couple of sample titles of research reports. "G-Protein Signaling from Activated Rat Frizzled-l to the BetaCatenin-Lef-Tcf Pathway." And, on the physical sciences side, this: "Chiral Sign Induction by Vortices during the Formation of Mesophases in Stirred Solutions." No acronyms there, at least. One might be tempted to ignore such stuff, but if you're serious about science, you can't; I assure you that these are important contributions in areas that one shouldn't just dismiss. This explosion of special knowledge has given many of us tough curriculum decisions. How do you decide what to teach and, harder still, what to leave out? How do you help students extract insight about the principles and structure of a science from what must seem like an endless parade of difficult and complex examples?

That is the intellectual piece of the problem. It is tough enough, but there is also an economic piece. I began to catch on to it in the 1980s as Stanford was recruiting and, if successful, appointing new faculty members in fields like chemistry and cell biology. At some point I recognized that the capital cost of renovating and equipping the appointee's laboratory and getting him or her set up was equal to, or greater than, the endowment needed to establish a permanent chair that would provide salary and benefits for the rest of the appointment! I decided that perhaps we had just encountered and crossed the divide that separates the watershed of "little science" from that of "big science."

So science, considered as an educational challenge, is migrating from something that can be circumscribed and contained to something so multiple and complex that it asks to be covered comprehensively, but at the same time defies being covered comprehensibly. And, on the economic side, it is becoming so method-intensive and so demanding of capital investments that it challenges the ability of institutions to pay for it.

Are there unique challenges facing liberal arts colleges tackling the challenge of offering integrated science curricula?

[The preceding] are problems enough for the research universities. They are magnified for the small liberal arts colleges, where smaller faculty size limits coverage of an increasing number of special areas, and where resource limitations constrain what can be done to support appointments in some of the most expensive ones. First, I pose a question that will seem a bit irrelevant, but needs an answer for other purposes, including the political: Why should we devote special attention to them? They do not constitute a major share of the nation's undergraduate enrollment, which is concentrated in the large public and private research universities and the comprehensive institutions. But the liberal arts college sector is a very significant contributor if one combines volume and quality. I remember that when I was a student, Swarthmore and Reed regularly appeared at the top of the nation's institutions in the proportion of their graduates who subsequently earned doctorates and entered academic careers. Indeed, in the period between 1935 and the end of World War II, the President's Scientific Research Board reported that four colleges in our present sample – Furman University, Oberlin and Reed Colleges, and Miami University of Ohio – had together graduated more students who later completed doctoral work in physics than Ohio State, Princeton, Stanford, and Yale combined.

What about now? I have watched Ph.D. students from the liberal arts colleges, superbly prepared, enter my own department of biology and then prosper. Today's numbers are even more impressive than the earlier ones I just cited. For the decade 1986 through 1995, the Ph.D. productivity of undergraduate institutions, baccalaureate-only colleges as well as research universities, was considered proportionately by comparing them with the number of bachelor's degrees each had granted in the previous decade. Eliminating the engineering-only schools and those that produced fewer than 200 graduates who earned Ph.D. [degrees] later, four of the top five institutions are liberal arts colleges. The top two, Reed and Swarthmore, nearly doubled the proportional productivity of Harvard and Yale. Even the absolute numbers contain some surprises.

So both in percentages and in absolute numbers, the liberal arts colleges supply an impressive percentage of the nation's doctoral candidates?

Carleton graduates over this period earned more Ph.D.s in chemistry than did those of Harvard, Yale, Stanford, or Princeton. And this trend is likely to continue. The liberal arts colleges are growing even more science-active. Enrolments in the sciences there, already substantially higher, percentage-wise, than in the research universities, are increasing as research support for their faculties has improved. Thus, there is reason to believe that the importance of the liberal arts colleges to future scientific power is, if anything, even greater than it has been. Why are these institutions so successful in this regard?

These colleges are able, by virtue of size and faculty/student ratio, to pay individual attention to the development of their students, to make sure that every science concentrator gets a full dose of laboratory instruction, and – in the absence of graduate students – to make undergraduate participation in original faculty research a regular occurrence. If their future as centers of science education is jeopardized, it will be a matter for serious national concern.

Are today's liberal arts colleges well-positioned to maintain their ability to deliver superb science pedagogy?

Had [you asked] a decade earlier, I would have started with an issue to which I devoted some space in a book I wrote in the mid-nineties, called *Academic Duty*. It is a supply-side problem: where are the prospective faculty members, and how are they being trained? The problem, as I saw it, had to do with the kind of education doctoral candidates were getting in the laboratories of our nation's research universities. That training, as splendid as it was in preparing them to do research of the kinds they had been doing as graduate students, prepared them for little else. Moreover, their mentors often left the impression that only a job like theirs, in an institution like the one in which they worked, would fulfill their expectations. As a result, relatively few Ph.D. candidates displayed a serious interest in settings in which they could not have graduate students of their own and thus, presumably, repeat the cycle.

Fifteen years ago when Peter Stanley was at the Ford Foundation, I remember discussions we had about how one might improve the chances that doctoral candidates in the sciences in such places might be persuaded to consider faculty positions at a liberal arts college. I don't remember that our enthusiasm for change came to much then. But serious efforts to change the patterns of graduate education are underway now, through the Pew program called Preparing Future Faculty, and in what George Walker, Lee Shulman, Tom Ehrlich and Chris Golde are doing at the Carnegie Foundation for the Advancement of Teaching. Lee speaks of changing graduate education so that its products feel responsibility for the "stewardship of the profession" including, of course, serious devotion to preparing the next generation.

And I do believe that things are changing in this regard. Not long ago I chaired a Ph.D. oral examination in chemistry; the candidate said afterward that he was headed for a liberal arts college by preference, and his adviser smiled proudly. Acquiescence would have been enough; this actually looked like approval. Better evidence comes from the testimony of liberal arts college presidents, who – because of the scale of their institutions – are really involved in the hiring process. The ones I have talked to don't put that problem at the top of their list of challenges.

What then are the most critical challenges facing integrated science education at liberal arts colleges and universities?

Strangely, this part of the story begins with some good news. Strong markets have stimulated donors everywhere, and raised endowments at the same time. A flush of new building construction has extended to the liberal arts colleges, and many of the institutions in the Academic Excellence Study sample report that they are better off in bricks and mortar than at any time in the past. The same growth has increased the number of endowed professorships.

This sounds auspicious – what is the downside?

The bad news comes in the form of two consequences that might have been expected. First, this has not been a rising tide that has lifted all boats equally. The long-lasting bull market may have benefited all institutions, but it has benefited some much more than others. Many good colleges with strong histories in undergraduate science training now see themselves as second- or third-tier, and may not be as relatively attractive to good students and faculty as they once were. (Of course it could be argued that concentration in the richest and most attractive institutions may be a public good - but that will be true only if the amount of training in the favored places increases enough to replace the losses elsewhere. I cannot escape the conviction that it will be a bad thing if the winner-takes-all phenomenon so evident in the rest of society comes to this domain as well.)

Second, the failure of equipment funding to keep pace with new building construction creates another odd disparity that is beginning to bother some science faculty. It is not difficult to think of one's splendid new building as "all dressed up with no place to go" when it is filled with outmoded equipment that looks shabby and out of place by comparison. Not only have the National Science Foundation and other funders not been sufficiently responsive; the cost of equipment has skyrocketed while useful product life cycles have contracted, trapping the liberal arts science departments in a relentless expenditure crunch – a down escalator from which some have simply had to get off. One of my interviewees described it this way: "What once were capital investments," he said, "have become consumables."

What about on the personnel side – how do these economic trends affect hiring?

There is some good news, though. Most liberal arts colleges are now able to appoint really good faculty. Partly that is because we are in a labor-excess academic economy, and of course that is not all good news. But that's only part of the explanation. Another is that endowment growth and enhanced giving not only built buildings; it endowed chairs, which donors like. Beyond economics, however, there is a more subtle incentive. A goodly proportion of the new appointees are young men and women who received their own undergraduate training in liberal arts colleges, liked it, and returned when the time came. Many, at least in my small sample, have returned to their own alma maters.

Do you view the migration of talent to liberal arts colleges as a "push" from research institutes or a "pull" to the liberal arts environment?

If we are to explain the phenomenon that a disproportionate share of the population entering the job market have come from liberal arts colleges, we cannot look only at the positive side of their experience. If the much larger, high-prestige research universities were sending their undergraduates on to doctoral study at the same rate, the pool would be much larger than it is. To explain why the undergraduates from the Stanfords, Berkeleys, Harvards, and Yales are entering academic doctoral programs in the sciences at a lower rate than Reed – not to forget Kalamazoo, College of Wooster, Earlham, Macalester and the other underendowed star performers – we can't ask only why the latter are doing so well. We have to ask why the former aren't doing better.

I wonder whether it isn't that the students in the research universities, as they observe life in those high-powered laboratories, and meet the graduate students and post-doctoral fellows, don't develop some reservations that make medical school look more attractive. They hear horror stories about the job market from the post-docs, and they hear graduate teaching assistants talking about a strike. Perhaps more important, they are able to observe the increasingly harried, committee-burdened lives of their faculty mentors and the narrow scope left for their family lives. It would not be surprising if some of those students, among them the very best, might conclude that this life just isn't for them.

Based on your observations, what are your recommendations for change?

That there is some bad news in this summary is not something to get seriously depressed about. The Academic Excellence Study . . . brings more encouragement about the liberal arts college sector. Faculty growth was accompanied by increases in public funding for research, and the publication record of faculty in the institutions sampled is surprisingly impressive: a peer reviewed paper every two years is nothing to sneeze at for a busy teacher. And the level of faculty research plainly has an influence, the study shows, on the probability that their students will move on to Ph.D. study.

This good news aside, however, there are needs - needs that must be considered if the liberal arts colleges are to continue their extraordinary historical record of success in preparing America's scientists. Plainly, their competitive advantage rests on the quality of intensive, individualized attention they are able to give to each student. Their disadvantage, if there is one, is that they lack some of the support mechanisms that are available in those research settings that are primarily oriented to graduate training. So here is a partial list of needs.

First, something I have heard time after time in talking with liberal arts college faculty and administrators is the need for funds to support summer research and fieldwork. Even where faculty members have grant support, as they do in unexpectedly large numbers, there is no provision for this important device, which is critical in enabling undergraduates to reach serious research depth. These days, students are devoting more time to work for pay during the school year, and are committing their summers to jobs that serve to decrease their debt loads. The loads themselves are, of course, powerful disincentives for engaging in graduate work. It is thus doubly important to give them opportunities that expand their futures as researchers, while not requiring them to sacrifice income.

Second, I found myself surprised at the degree to which the facilities problem, so pressing in earlier times, has now become an equipment problem. Some of the funding agencies are beginning to recognize this; but for all but the richest of the colleges, laboratories are under equipped. (I note that this problem also affects the top-of-the-line research universities: the equipment available in the best industrial research laboratories is far better than what was available to their young scientists in the university labs they left.) As for the liberal arts colleges, since the strength of science education there lies in shared research experiences between faculty mentors and bright students, the equipment problem – and its impact on the cost of setting up new science faculty – is a critical bottleneck. Relief ought to be a high priority for funding at the National Science Foundation and other agencies.

Third, there is still a problem in the preparation of future faculty to serve in liberal arts college settings. Too many doctoral candidates and post-doctoral fellows in the research universities are still not effectively mentored in small-group teaching, and too many are left with the notion that the colleges are not suitable places in which to pursue a serious scientific career. One hopes that various projects for reforming doctoral education will remedy this situation, but more focused attention from supporting federal agencies and private funders may be necessary to accelerate change.

Chapter 7
Do the New Directions in Scientific Training Have an Impact on Developing a More Diverse Workforce?

Rita R. Colwell, University of Maryland and 11th Director of the National Science Foundation

As Director of NSF, Dr. Colwell was at the forefront of challenging institutions to increase the diversity of those engaged in scientific enterprises.

M.E. Brint et al. (eds.), *Integrated Science*, DOI 10.1007/978-0-387-84853-2_7,
© Springer Science+Business Media, LLC 2009

Could you begin with some personal history? As a woman, did you find obstacles to fulfilling your goal of becoming a scientist?

...When I went to high school, girls simply were not allowed to take physics. More to the point, my high school *chemistry* teacher told me I would never make it in chemistry – because women could not. Some attitudes, we can be thankful, have been exposed as patently absurd; some "rules" have been rewritten. Today we rarely hear blatant condemnation of women in science. Women have entered chemistry in droves, and now receive about a third of the Ph.Ds.

While many women chemists, from Mary Good to Mary Anne Fox, from Elsa Reichmanis to Helen Free, currently play visible and superb leadership roles, we all recognize that we still have a long way to go to create the diverse workforce our nation needs. The number of women receiving chemistry doctorates has risen steadily, but not women on senior faculty of our colleges and universities. It's high time we rethought some other rules, those that govern women's advancement in academia. As Princeton chemist George McLendon put it, "...Academic institutions are intrinsically monastic institutions that were created in the 13th century. They might need a little fine-tuning."

Jumping ahead a few centuries, do you think that there are reasons to be optimistic about the success of diversity in the academy today?

[Let me] note a few reasons why science needs diversity; then move to a broad sketch of women in science generally and in chemistry in particular. I'll emphasize where the real rethinking of rules must take place – within institutions, and then cite examples of such activities. For a microbiologist like myself, there are some truths about the value of diversity and environment in natural systems that are tempting to apply to human systems as well. The interconnectedness of life is a very deep law, and greater diversity makes for a more robust ecosystem than does a monoculture. The environment must nourish any organism, or it will not survive – just like the environment for a young scientist, which can be chilling or nurturing.

Another truth of the natural world is that systems are always evolving. Indeed, the vitality of the ecosystem depends on this ability to adapt and on its complexity. Our nation is changing. As our economy becomes more knowledge-driven, we will need to draw on the more-than-half of our population that avoids the pursuit of science and engineering, or we will not be able to maintain the pace of discovery. As our national workforce becomes increasingly diverse, a scientific enterprise with mainly white, male faces risks sending the signal that others are not welcome.

Who teaches the next generation is important. One obstacle to increasing diversity is what's been termed "the reflecting pool" – the tendency for faculty to hire and promote those like themselves. At the teaching level, the percentage of women faculty has been cited as the best predictor of future success of female undergraduates. After my negative experiences with teachers and counselors in high school, college certainly offered me an outstanding mentor – Dr. Dorothy Powelson, a bacteriologist who inspired a number of women, myself included, to choose science as a career.

Diversity propagates diversity. Faculty demographics that reflect the student mix will enable better mentoring. A diverse faculty also serves as a beacon to young scientists who are exploring the possibilities for their futures. For a moment, let's step back for a wider inspection of where women stand in higher education. Of senior faculty in natural science and engineering, just 12.5 percent are women. Overall in science and engineering, 94 percent of full professors are white, and 90 percent are male. Furthermore, women earn less at every level. The latest data from 2000 show women earning 36 percent of science and engineering Ph.Ds, from 16 percent in engineering to 67 percent in psychology. Still, the problem for women in many fields is largely not in getting a Ph.D; it is their differential status after they enter academia.

Where does chemistry stand for women, compared to other sciences? At the National Research Council's 2000 workshop, "Women in the Chemical Workforce," Margaret Rossiter, Cornell University historian of science, said, "Many fields of science are doing better than chemistry – including many but not all of the social sciences and biology. Chemistry departments are just now getting to the point where almost all have one or more women faculty members."

While that may look like progress, one or two women in a department of 15 to 20 faculty is hardly a profound change. If women are only beginning to breach the glass ceiling in academic chemistry, underrepresented minorities have an even longer way to go – according to Donna Nelson's celebrated study of chemistry faculties. Nelson, of the University of Oklahoma, surveyed the top 50 chemistry departments. Of more than 1600 top faculty members, only 43 were minorities. Perhaps her most striking finding was that chemistry departments of the top 50 had "zero" African-American assistant professors – even though 319 chemistry Ph.Ds were granted to African-Americans between 1991 and 1999. The number granted per year has also risen, from 23 in 1991 to 56 in 1999. In addition, there were only 17 African-American faculty in total – 11 full professors and 6 associate professors.

Women, too, are lost from academe's "leaky pipeline" every step of the way, all the way up to the senior faculty level. In economic terms alone, that's a disastrous investment strategy. At the top, in the National Academies of Science and Engineering, whose membership is one of the highest honors in science, there are only a handful of women chemists, and even fewer African-Americans. Between 1923 and 1970, ten women of any discipline were elected to NAS.

Currently, the National Academy of Sciences has 154 women out of a total of 2304 members, about 6 percent. At the National Academy of Engineering, there are 67 women out of 2233 – 3 percent. At the Institute of Medicine, women number 263 out of 1404 members – 19 percent – 16 percent if you don't count *emeritus* and foreign members. The National Academies of Science and Engineering do not officially track the race of members. The IOM does: it has 82 African-American members, 1 Native American, and 17 Hispanics. To be sure, the situation has been improving. For example, female members of NAS have doubled over the past decade. However, women NAS members are not well-represented across fields, with the majority in the biosciences.

At the other end of the spectrum, at beginning career levels, many women scientists never enter academia, or leave it in discouragement, after finding the environment isolating, cold, and hostile. It has been observed that more women chemists seem to be choosing industry. . . .NSF chemist Marge Cavanaugh observes, "I think there is a push away from universities as well as a pull by industry." It's possible that industry offers a better environment for women to succeed – with more-defined rules, rewards for hard work, including better salaries, a reasonable workday, a clear bottom line, and more control over one's life.

What factors present obstacles to the rise of women who do enter academe?

We are realizing that we must foster change not in individual women – remember the growing numbers of women Ph.Ds – but at the institutional level. As Virginia Valian of Hunter College observes, ". . .Women start out slightly behind men in rank and tenure and become increasingly disadvantaged with age." We have learned that subtle, small disadvantages accumulate to hold back women scientists and engineers. As a computer simulation has shown, even a tiny bias towards promoting men can result in an eventual domination of a hierarchy by male leaders.

. . .Data has shown that less lab space, research time, travel funds, and all the small perks that make a huge difference over a lifetime can thwart even the most hard-working and dedicated woman scientist and engineer. . . .Certainly much thinking should take place about rewriting the rule of balance between family and career in academia, a balance that is difficult for both men and women, but that often proves more complex for women. Many factors foster success – there is no one way to balance the equation. We have become very aware, however, that academic institutions must be partners in creating environments that truly value diversity. As an example, the Massachusetts Institute of Technology has done yeoman's work in examining the situation of its women faculty.

One striking finding a few years ago was that no woman in the School of Science had ever served as department head or director of a center or laboratory in MIT's history. Initially, this and other problems were addressed on a case-by-case basis. Now, problems are dealt with on an institutional level. "This is a profound change," says a new MIT report, "probably the most important to occur for some decades."

There is new realization that the time-honored system of compensating faculty may be "gender-biased" – favoring men but discouraging women or two-career couples. After its study of science faculty, MIT expanded the investigation to other schools. At the management school, men and women's experiences were discovered to be so different that they were, in effect, working at different schools – a finding that Dean Richard Schmalensee termed "profoundly disturbing." In the engineering school, Dean Thomas Magnanti wrote in a statement, "We learn [that] some of our women faculty . . .have never been asked to serve on the Ph.D. committee of even one of their colleagues' students . . .stunning." MIT's former Provost, Robert A. Brown, has noted some institutional changes that could certainly help improve the academic climate, ranging from identifying a larger number of women and minority candidates for positions, to delaying the tenure decision by one year for a woman having a child, to arranging a half-time appointment for faculty caring for a family member.

We can see how such changes may indeed benefit not only women but everyone involved. Such changes, we hope, will also make academic careers more appealing to young people and halt the current squandering of science and engineering talent. At a fundamental level, these changes must be embraced at the very top of universities and embedded in university policy.

Diversity gives greater scope for adaptation and innovation – traits our social systems, our nation and our economy, also need. A more diverse science and engineering workforce will bring in different talents, approaches and experiences. Diversity itself becomes a critical attribute, not to mention flexibility, innovation, and creativity. All are factors for success, especially now, for doing science and engineering in the [21]st century.

Donald Kennedy, President Emeritus, Stanford University

Educated at Harvard University, Dr. Kennedy has spent most of his career at Stanford University as a faculty member, Provost, and President. From 2000–2007, Dr. Kennedy was editor-in-chief of *Science*, the prestigious magazine published by the AAAS.

Dr. Caldwell has advocated changes that would diversify the pool of scientists at the university level. How effective have these efforts been?

The disappointing percentage of scientists – especially science faculty members – who are women and members of minority groups has been a source of anxiety for years. In response, universities and other employers have initiated heavy recruitment efforts to improve their performance. The results offer some encouragement: the increase in the proportion of tenure track faculty who are women, . . .is certainly a good sign. Yet in many respects the process has looked more like reshuffling than real reform. The problem is that the pool isn't increasing fast enough.

You have spoken of the role of liberal arts education in relation to science research and education. What role do or should liberal arts colleges play in developing an environment that encourages diversity among scientists?

The liberal arts colleges may be uniquely positioned to do something about that. Many of the complaints about the discouraging climate faced by members of minority group members and women speak to the impersonality of science instruction. The negative effects on women are well illustrated in "Talking About Leaving", the persuasive account put together by Seymour and Hewitt at the University of Colorado. In it, women speak in their own voices about their experiences. They are often explicit about their need for more individual mentoring and more personal encouragement, both about their performance as students and researchers, and about their later careers. Those needs, shared by members of all minority groups, play directly into the long suit of the liberal arts colleges. Evidence that it works is found in the strong records of women from such institutions, including some that enroll women only. It is also found in the graduates of some of the smaller, historically black colleges and universities. I think particularly of the remarkable record achieved at Xavier University in New Orleans by Norman Francis and his colleagues.

I wonder whether a national effort to support and promote this activity in these places would not have a greater impact than all the resources now being spent on a game that often amounts to little more than musical chairs. The last simile may seem a bit harsh; after all, the highly visible hires do enhance the national sense that women and minorities may succeed at the highest levels, and that should help morale. But we are unlikely to have a systemic, long-lasting improvement in faculty diversity unless we can enlarge the size of the pool. That means making science attractive and fun, and doing it person-to-person. Who is in a better position to do that than the professoriate in the liberal arts colleges?

Chapter 8
What Are the Implications for Training at the Master's Level?

Rita R. Colwell, University of Maryland and 11th Director of the National Science Foundation

As Director of NSF, Dr. Colwell was at the forefront of challenging institutions to increase the diversity of those engaged in scientific enterprises.

M.E. Brint et al. (eds.), *Integrated Science*, DOI 10.1007/978-0-387-84853-2_8,
© Springer Science+Business Media, LLC 2009

Dr. Colwell, could you provide a general overview of the changing context for graduate education and training as you see it? [1]

In recent years, interdisciplinary and integrative research has become synonymous with all things modern and progressive about scientific research. This new imperative has not been based in some simple philosophic belief in "heterogeneity" but in the complex nature of current scientific and societal problems. It is argued that, in many fields, the easy work is finished and ambitious scholars are confronted with problems that not only defy the specialization of disciplinary skills, theories, and methods but actually demand their collaboration. As a result, academic science – and perhaps knowledge production more generally – has engaged in a transformation from the "old" way of doing research to a "new" way. Whereas the former is characterized as homogeneous, disciplinary, hierarchical, and permanent; the latter is heterogeneous, interdisciplinary, horizontal, and fluid.

While such changing circumstances can be unsettling to those involved, science and its practitioners should not fear that the rising tide of interdisciplinary methods will displace traditional disciplinary harbors. To the contrary, disciplinary bases will remain essential, both in terms of their contributions to knowledge and as foundations of learning, at the same time that interdisciplinary crossings will become crucial in the reconfigurations and applications of such knowledge and learning. Thus, the challenge for universities is not only to transform the way they do research currently but to reformulate the way they prepare disciplinary practitioners, as well as interdisciplinary pioneers, to do research in the future.

At the same time that these organizational and epistemological shifts are happening in knowledge production, structural and professional changes are also taking place in the labor market. Traditionally, science and engineering graduates have been prepared for positions in academia, industry, and government laboratories where scholarship and research – especially "basic" research – constituted the primary focus of employment. However, a persistent long-term trend away from posts of "basic" research and teaching in the academy toward positions of "applied" research and non-university employment has meant that less than 50 percent of the total science and engineering Ph.D.s have been employed in academia in recent years. Yet, as more doctoral recipients are either seeking out or turning to employment in 2- and 4-year colleges, non-profit and corporate businesses, and governmental agencies, their employers are complaining that new Ph.D.s are trained too narrowly to manage the range of professional tasks they encounter.

Thus, beginning with the 1995 COSEPUP[2] report Reshaping the Graduate Education of Scientists and Engineers, a call has been made to reform graduate education and training programs in science and engineering. Programs have been

---

[1] Rita Colwell, "The Changing Context for Graduate Education and Training," taylorandfrancis. metapress.com/index/WJ47M18Q6836V335.pdf

[2] Editor's Note: COSEPUP is the Committee on Science, Engineering and Public Policy of the U.S. National Academy of Sciences.

encouraged to retain the strengths of the existing graduate education and training system but in a manner that better integrates education with research and further incorporates "hands-on" practical experiences so as to expand the intellectual versatility and practical applicability of future scientists and engineers who may work in either basic or applied research settings.

Given your analysis, what are the implications for the design and delivery of graduate education and training?

As a result of such organizational and epistemological changes in knowledge production and accompanying structural and professional shifts in the labor market, efforts are underway to reform graduate education and training programs in ways that prepare students for these new models of scientific research and these new modes of scientific employment. Together, these contextual changes and reform efforts have led to the development and diffusion of what Diana Rhoten and Edward Hackett call "innovative, interdisciplinary, and integrative" – or I3 – approaches to graduate education and training. Such I3 programs seek to: (a) ground students in the fundamentals of their own fields as well as expose them to several subfields of science and engineering; (b) develop students' technical proficiencies as well as their abilities to communicate complex ideas and to work well in teams; and, (c) prepare students to engage the diverse publics concerned with science and technology in ways that shape policy and inform practice in various sectors and contexts.

The California State University (CSU) system is exploring the feasibility of implementing a new class of master's degrees systemwide, in a bid to boost the state's science and technology workforce. Professional science master's (PSM) degree programs have been steadily growing nationwide over the past decade, but this is the first time a state university has considered implementing the programs on a system-wide basis. PSM programs differ from typical science master's degree programs in that they attempt to better prepare students for employment in the business environment, usually by incorporating business coursework into a more traditional science curriculum. The programs are generally open to bachelor's degree holders in the sciences, mathematics, or engineering. These programs consist of two years of training in an emerging or interdisciplinary area. Many include internships and "cross-training" in business and communications. The Alfred P. Sloan Foundation initiated a program in 1997 to promote development of these programs, and has since provided over $12 million in seed grants at programs throughout the country. Today, largely as a result of the Sloan initiative, PSM degrees are offered in at least 45 institutions around the country, including UC Los Angeles, the University of Southern California, Stanford University, CSU Fresno, and the Keck Graduate Institute of Applied Life Sciences. Subject areas range from biosciences management to physics for business applications. However, these programs are not numerous or large enough to produce a significant number of graduates; yet only 400 PSM graduates have been produced nationwide since 2000 (compared to over 24,000 S&E master's degrees issued during the same period in California alone).

Sheila Tobias, National Outreach Coordinator for the Professional Science Master's (1997–2005)

In addition to authoring eleven books on subjects relating to women and science, mathematics anxiety, and science education, Sheila Tobias has been a faculty member and university administrator.

As a way of defining the Professional Science Master's degree mentioned by Dr. Colwell, let's begin with a passage from a recent report by The California Council on Science and Technology that you helped craft:[3]

PSM programs are designed to prepare graduates for entry-level professional positions in business, government, and non-profit employment sectors. PSM programs generally consist of a core of advanced work in the discipline or an interdisciplinary area, augmented by work in information technology and related disciplines. They also include short courses or modules that expose students to fundamental elements

---

[3] "An Industry Perspective of the Professional Science Master's Degree in California," *California Council on Science and Technology*, January, 2005, pp. 1–2 and 15.

of business environments, such as business basics, intellectual property law, team-work, communication/presentation skills, and project management.

Most PSM programs are created by combining existing science and business coursework offered by the institution. A few institutions, such as the Keck Graduate Institute of Life Sciences, have relied primarily on coursework designed specifically for the PSM program. By making use of existing resources and faculty, the Sloan approach is designed to be less expensive to initiate, requiring an initial investment of no more than $75,000–$300,000 to seed a program. As of 2007, approximately 1,500 PSM graduates have been produced nationwide.

Could you give some background on the creation of the Professional Science Master's Degree that you helped to develop?

In 1995, the Committee on Science, Engineering, and Public Policy (COSEPUP) of the National Academy of Sciences suggested that graduate schools of arts and sciences consider a "different" kind of graduate degree, less oriented toward research and requiring less time to obtain. The COSEPUP report was a long-delayed acknowledgement that graduate education in the sciences (and mathematics) had become severely decoupled both from the career needs of students and from the supply needs of an increasingly technology-based national economy.[4]

Even before the National Academies had broached the topic, there were stirrings in the mathematics and science communities. The Society for Industrial and Applied Mathematics (SIAM) was at the helm of the nascent movement in mathematics. SIAM recognized early that not only were there jobs to be had and interesting careers to be made by mathematicians in business and industry, but that some of the more interesting problems in mathematics could come from such settings.[5] In contrast, for most of the sciences reform was thought about in terms of improvements in doctoral education.

All the while, the "silent success" of the terminal (professional) master's degree in fields other than science and mathematics was pointing in another direction. Beginning in the late 1980s, an increasing number and proportion of M.S. degrees were being awarded in professional fields, with M.S. degrees in arts and sciences in decline.[6] In a 1995 study, "Rethinking Science as a Career," my collaborators, Daryl Chubin and Kevin Aylesworth, and I asked the question: "What kind of professional master's degrees might we invent for science?"[7] And, once these degrees were in

---

[4] Sheila Tobias, "The Professional Science (Math) Master's Degrees: History and Prospects," The Communicator, Council of Graduate Schools, www.cgsnet.org, Volume 39, Number 6, July 2006, pp. 3–6.

[5] *The SIAM Report on Mathematics in Industry*, SIAM Publications, Philadelphia PA, 1995.

[6] C.F. Conrad, J.G. Haworth, and S. B. Millar, *A Silent Success: Master's Education in the United States* Baltimore: Johns Hopkins University, 1993.

[7] Sheila Tobias, Daryl E. Chubin, Kevin Aylesworth, *Rethinking Science as a Career* (Tucson, Arizona: Research Corp., 1995).

place, could they be marketed to faculty, students, employers and the public more generally? If one looks to the master's degree in fields other than science and mathematics, one finds that instead of training producers of scholarship – the traditional purpose of graduate education – master's educators aim to produce people who are able to use the products of scholarship in their work and who are familiar with the practical aspects of emerging problem areas. The general outlines of a professional science or mathematics master's degree were available to the community by the mid-1990s.

Who were the students you were targeting for these degrees?

Beginning in the 1990s, concern began to be expressed as to the underenrollment of university students in physics. In the U.S., where university students can switch in and out of majors throughout their course of study, failure to recruit is compounded by failure to retain. U.S. colleges and universities have lost 16 percent of their first-degree physics graduates in the last five years (24 percent in the last 10 years), dropping from a nationwide annual graduating cohort of 4,600 to 3,800 – a 40-year low – compared to the production annually of 78,000 engineers.

In fact, fewer than one-third of U.S. physics majors actually become "physicists" (as defined by the profession). Research findings indicate that students do not choose science in general, physics in particular, because of "narrowness of study" and "inflexibility" as regards future employment. Given these findings, perhaps a physics curriculum designed to prepare students for further physics study is not the most appropriate curriculum for the rest. Not just the teaching, textbook, and laboratory, but the physics curricula are being increasingly subjected to review. Supported by national funding and by one another, we now have a large number of physicists – even Centers for Physics Education Research – dealing with the "best practices" in the teaching of physics and the training of physics instructors, as well as the productive uses of "project-based teaching," "workshop physics," and "peer instruction."

But is this enough? Will improved pedagogy be sufficiently attractive to win back the physics student who likes and does well at physics, but isn't inspired by the career options currently available? Some of us who care both about the continued health of physics and about the lack of physics-trained professionals populating other fields (e.g., banks, foundations, government) think that, in addition to improvements in teaching, some altogether new graduate degree programs are called for, leading to careers related to physics but not circumscribed by physics and its immediate applications. Those involved in these new degree programs are calling them "Professional Masters" intending therewith to convey their equivalence (if not congruence) with the MBA and LLD degrees.

It is easier by far to create new programs within existing degree structures in higher education than to create new degrees. New programs certainly have their challenges. They often involve wholly new techniques or new applications of old techniques. They challenge the notion of who is the "expert" in a changing field, one in which no "expert" received his or her degree. And quality control has to be imaginatively reinvented when the lines between disciplines are blurred. Finally, there is the issue of pedagogy: how is one to teach a new subject?

Though often stimulated by the pressure of emerging fields, the creation of new degrees is not an inevitable consequence of new discoveries or new research directions. In my view, a new degree has to have its own dynamic and rationale to be supported by higher education administrators and their stakeholders. Thus, the new professional M.S. degree in the sciences (and mathematics) is not to be construed as just a way of accrediting or institutionalizing or allocating scarce resources to one or more emerging fields. Rather, it is a means of providing and legitimizing some post-baccalaureate alternatives for science and mathematics majors who don't wish to do medicine or engineering, or pursue a research Ph.D., but do not wish to leave science altogether.

Until now, these students have had nowhere to go, except directly into industry where, as terminal BS or BA degree holders they are housed in research labs as techies. There is of course the MBA or law option. But for many of the students we are eager to serve, such post-graduate options cause them to have to reinvent themselves as law or business students, and to compete with students who, while not well-schooled in mathematics or science, are able to compete with them in the arts of advocacy and/or marketing.

Are there any analyses that look at the effectiveness of these new Master's programs?

It wasn't until 2004–05 that serious analytic studies of the professional master's in the sciences and mathematics began to appear. Judith Raymo who had previously written about both master's education, and the doctor of arts degree, published *Professionalizing Graduate Education: The Master's Degree in the Market Place*.[8] Les Sims, a chemist and former graduate dean at the University of Iowa, worked on the CGS/Sloan initiative. His book, *Professional Master's Education: A CGS Guide to Establishing Programs*, provides a comprehensive outline of best practices, activities, and processes for professional master's programs.[9] Raymo locates the new degree at the intersection of two innovations: that of interdisciplinary science (e.g., bioinformatics, nanotechnology, robotics, systems biology) and that of applied tech-business-oriented training. Les Sims, too, sees greater benefits even than workforce enhancement in the propagation of the professional science master's. "Such post-graduate options," he writes, "have the potential to encourage a larger percentage of students to pursue graduate education in the field of their major."

---

[8] Judith Raymo, *Professionalizing Graduate Education: The Master's Degree in the Marketplace* (San Francisco: Jossey Bass, 2005).

[9] Leslie B. Sims, *Professional Master's Education: A CGS Guide to Establishing Programs* (CGS, 2006) *by Sheila Tobias, Consultant, Council of Graduate Schools.*

What is your hope for the future of the PSM program?

Until 2007, the Sloan Foundation – along with the Keck Foundation, the mathematics community, and local initiatives at a few dozen universities – have supported the new professional master's degree with hopes of positively impacting the science/math pipeline in a number of ways: seed reform in graduate education more generally; keep non-research oriented students in the major; provide a cadre of science-trained professionals useful to business, industry and government; and bring more science-trained people into the power centers which tend to be over-populated with professionals from the law and finance communities.

Then, in the summer of 2007, a section was added to the America Competes Act, endorsing the PSM as part of a wide-ranging bill focused on higher education, discovery, and innovation. At this time, the final appropriation is still uncertain, but some $10 million for the first year, $12 million for the second year, and $15 million for the third year was authorized. As important, the National Science Foundation, which until 2007 had shown interest in but no willingness to support the effort, is obliged to set up a Clearing House of Information about the PSM.

Since the professional MS in science and mathematics is characterized not only by new subject matter but also by new pedagogies and new means of evaluating applicants for admission and matriculation, PSM programs are not identical. ...Some involve an emerging new field, such as bioinformatics; others interdisciplinary study, such as computational sciences; still others, science/mathematics "plus" business, law, organizational theory, and communication. New pedagogies are also being discussed, such as case studies within a modular schedule at the Keck Graduate Institute; a lab rotation through a variety of cutting-edge research fields at the University of Arizona; and, at Michigan State University, a series of "basics for business" weekend short courses tailor made for the background and mindset of science/mathematics
students.

Like the MBA, which took nearly 40 years to sell to students and employers, these programs must be packaged and sold. What we are after is a high-level education in the science/mathematics underpinnings of today's and tomorrow's technologies, one that will offer graduates flexible careers at the interface of R&D, product development, regulatory affairs, intellectual property issues, marketing, finance, and management. Let us hope students will perceive these opportunities the way we do and that employers will provide innovative career pathways once they are in the work force.

Chapter 9
What Are the Implications for Training at the Doctoral Level?

David Baltimore, Robert A. Millikan Professor of Biology at California Institute of Technology

Dr. Baltimore, Nobel Laureate, received his doctorate from the Rockefeller Institute. He served as President of the California Institute of Technology from 1997 to 2006, when he was elected to a three-year term as President of the American Association for the Advancement of Science. The research of Baltimore, Temin, and Dulbecco helped illuminate the role of viruses in cancer; the three men shared a Nobel Prize in 1975.

M.E. Brint et al. (eds.), *Integrated Science*, DOI 10.1007/978-0-387-84853-2_9,   109
© Springer Science+Business Media, LLC 2009

Dr. Baltimore, could you put into historical perspective the interdisciplinary changes that have recently occurred in biomedical research?

[Let's] consider what's happened in the last few decades. I will focus on basic science first because that is where the new advances are developed that ultimately affect clinical medicine. The huge step forward was the development of the vision of biology as an information science. With the discovery of the structure of DNA, we saw the 4 bases that provide sequence in DNA and it was evident that this sequence represented a code. The code had four letters, not 2 as in computer codes, but was in all respects a true code. Thus, in 1953, biology became a science of information even though we called it molecular biology. The molecular biology that was born in the 1950s, flowered in the 1960s and 1970s and has flourished thereafter. It arose out of a very simple experimental paradigm. There was a cadre of mainly academic scientists who ran small laboratories that were at once sites of training for the next generations and sites of discovery. There were almost no pure theoreticians who mattered but the experimentalists doubled as theorists, letting their theory direct their experiments and their experiments drive their theory. Even Francis Crick, the purist theoretician, did a few experiments. Two of the deepest theoreticians, Francois Jacob and Jacques Monod, were very productive experimentalists and until recently Jacob was still running a laboratory in Paris.

From the 1950s to the 1990s, this paradigm held sway. When a new technique was born in one of these little labs, it was quickly spread by the post-docs and graduate students because it generally required little sophistication and relatively simple apparatus. In parallel with this cottage industry of highly productive science, there arose a small industry, led by Arnold Beckman, which provided logistic support, producing new and powerful instruments but generally ones that could be run by the technologically challenged. At the same time, there were laboratories that ran differently because they did a different kind of science: these labs were involved in fields like neurophysiology and structural biology which required an understanding of electronics, optical theory, and other technological skills. They turned out to be harbingers of the future.

In the 1990s the change began. Actually the roots were in the 1980s where Lee Hood at Caltech had a different vision, one that involved building sophisticated instruments as an integral part of the experimental enterprise in molecular biology. That meant getting people into the biology laboratory with new skills, like machining, computer programming, and optics. Hood's operation at Caltech produced among other instruments, the gas phase protein sequencer and the automated DNA sequencer. It was the DNA sequencer that produced the deepest revolution in biological science.

The sequencer provided the technological leap that gave birth to the Human Genome Project. This project, so successful today, required a different sort of laboratory, one with a more industrial feel. Some of biology became like high energy physics, with large, interacting laboratories with powerful leaders producing a volume of data never before seen by biologists. The data spawned a field of computational biology, really a type of theoretical biology driven by the huge quantities of data suddenly spewing forth from these data factories.

This changed paradigm of research is having deep ramifications in biology. Even the older style small scale laboratory scientists can't ignore the power of the new methods and all top universities are scrambling to adapt, often by developing new types of departments and new levels of support facilities.

What are the implications of this new paradigm with respect to scientific training?

To answer this, we need to project ahead and think about the structure of biological science in the next decades. So I will peer into my crystal ball and make some suggestions about what may happen. But let me do this by analogy. I suggested that biological research is moving toward the patterns already established in physics. How does physics work? There are actually three major kinds of physicists: theoreticians, experimentalists working in high energy physics and small-scale physicists working in solid-state physics. There is also, of course, astrophysics, which is a hybrid field at the moment. But I want to focus on the examples of high energy physics. There is a clear division between the experimentalists who work in huge groups, often at accelerators, and theoreticians who stay at universities. They play equally important roles, although theory recently has gotten very far away from data, a worrying trend because classically experimental data both drove theory and checked theory.

I see biology moving towards a 3-tier structure. There will be the traditional small-scale laboratories, supported by shared facilities, there will be the large-scale experimentalists and there will be the theoreticians. There may never be the total split between theory and experiment that we see in physics, but the split will be there. And like what is true in physics, the most creative people, the smartest people, will be in theory. They will drive the experimental directions but not actually do the experiments. Their training will be in biology but also in physics, chemistry, and applied mathematics because they will have to be able to handle and integrate a wide range of physical, chemical, and systems issues.

How about those people who are still running the old-fashioned style of laboratory, will they be left behind?

For the next few decades, I am sure they will continue to play important roles. After all, for all the noise about the Genome Project, where have the really innovative ideas come from in the last decade? The Genome Project has not actually given us many new concepts, it has mostly posed questions for the future and been a hand-maiden of small-scale work. Notions like prions or interfering RNAs, which have added whole new ideas to the thinking of how biological systems function, have come from the traditional laboratories. The whole revolution in cancer research was fueled by individual investigators working with their graduate students and post-docs. The enormous need we have to understand autoimmunity may be helped by genome screening techniques but I expect that progress will largely come from individual immunologists working out the roles of cytokines, chemokines, and intra-cellular mediators of response.

With this background in mind, how do you prepare people for this future? If there are multiple styles of research, should there be multiple styles of training?

I think the answer is yes. We need people trained in a wide range of disciplines to be our experimentalists. They need to know biology but also need to know how to build machines, how to manage large numbers of people, how to get money on a huge scale. Our small-scale researchers need more training than ever in engineering, mathematics, and physics to be able to bring to bear new technology on problems. But it is the theoreticians who need the widest range of skills because they have to integrate all of science and focus it on biology. They have to be able to access complex data in huge databases, to manipulate it and to find new meaning in it. They will be trained in the best centers but in the way of theoreticians, they will not be disciples so much as *sui generis*. They will need powerful institutions in which to work but they will have to be independent of the institutional structures, able to roam freely, find the training they need wherever it is located, and produce idiosyncratic integrations.

What institutional changes will be required to meet these training needs? How does our present-day educational and grant structure map onto these needs?

Not particularly well. Almost all of our training structures depend on a senior investigator who generally produces people in his or her own image. Most of our training is in traditional laboratories. Most of our trainees have narrow backgrounds in biology and have not been educated in mathematics, physics, or engineering at the levels they will need. Most importantly, we have a system that emphasizes dependence, not independence. Professors have tight control over their trainees and hold their future in their hands. I am still writing letters for people who graduated from my laboratory 30 years ago. Today I warn my trainees that I am not preparing them for the future; I see ahead of them. I always hope that they will go in directions that surprise me, where I can't necessarily help them at all. But I am in the privileged position of being able to encourage independence; most principal investigators have a great need for their trainees to contribute to their own reputations. One of the worst aspects of this is that trainees are kept in training too long.

Having gotten to the issue of the length of training, let me spend some time on it because I consider it one of the key issues of the present time in biological science. There are many dynamics leading to trainees spending much too long in a dependent position where they do not have freedom to develop and test their own ideas and directions in science. Let me enumerate them:

1. As I have already mentioned, it is valuable for principal investigators to keep trainees for a long time because they become more skilled and more productive over time.

2. The second dynamic at work is that biology has become much more complicated to do. This requires developing a wider range of knowledge and that takes time. Also, important work often integrates concepts from multiple areas of science, requiring a trainee to learn, and work in multiple fields.
3. The third dynamic is that the trainee is expected to have both graduate training and post-doctoral training. This is not only important for credentialing but it is also a reality that it takes a long time to develop all the skills needed to run a complex laboratory. And then many trainees decide that the field of their graduate work is not where they want to make their career.
4. The fourth concern is that once a trainee lands an assistant professorship, she has little choice but to continue working along the lines established in her mentor's laboratory because the tenure system means she will be judged for tenure within usually 6 years of arriving. Six years is not enough time for her to develop an independent line of work and reach the international reputation needed at a good research university to receive tenure. Some schools have a longer road to tenure, which does give the investigator a chance to do something new but at a great risk.
5. The M.D. or M.D./Ph.D. has an even longer period of training. Not only must he or she get an M.D., but if the M.D. is to be valuable, there is further training to do. I know of young Assistant Professors in their 40s.

So we have a series of dynamics at work that mean that an investigator is not truly independent until the age of about 40 and for many, especially M.D.-Ph.Ds, it may be significantly later. I don't want to get into the issue of whether scientists are most creative in their 20s and 30s, although I suspect that is true. I merely want to point out the 40s is when you are supposed to have a mid-life crisis rather than just be starting out on a career. Also, it is not just that we are potentially losing the most creative years of scientists; the long period of dependency is also a disincentive for a budding biologist to enter the field or for a scientifically minded M.D. to pursue research.

I could also point out that when we at Caltech hire a theoretical physicist or a mathematician, the successful candidates are often in their late 20s and generally achieve tenure in their early 30s. I truly consider that the situation in biology today is a crisis and put into the context of our need for creative people in the next generations it is a disaster.

What can we do about this? We must break the cycles of dependence. The effectiveness of laboratories should not depend solely on trainees, we need full-time researchers in them who have made a career of research. It must be recognized that this will be expensive. We need to treat such researchers as professionals, giving them living wages and full benefits. We need to help trainees develop faster by focusing on the factors that work against this. We particularly need programs that encourage independence. I started one at the Whitehead Institute 20 years ago that has been notably successful in producing young, independent investigators. I need only mention Eric Lander, Peter Kim, and David Page to illustrate my point.

[Let me] end by reiterating that no matter what happens to the structures of science, creative, devoted people are the key to progress. We should never lose sight of that perspective. We should build our funding structures, our institutional structures, our evaluation structures around optimizing the ability of the best people to express their creativity, and rewarding those who do. We should be conscious of when our structures have started working against the ideal and adjust them.

To summarize, the successes of biology have established new challenges for the training process. Meanwhile, the present system has some serious defects. Put these two together and there seems little doubt, at least to me, that this is an issue that needs some serious thought.

Chris M. Golde, Associate Vice-Provost for Graduate Education, Stanford University

Dr. Golde received her Ph.D. degree from Stanford University in 1996. Her contribution was composed while she was an Assistant Professor at the University of Wisconsin-Madison.

H. Alix Gallagher, Senior Researcher, SRI International

Dr. Gallagher received her Ph.D. degree from the University of Wisconsin-Madison in 2002.

Dr. Baltimore has spoken of the need to reform the structure of graduate education. Drs. Golde and Gallagher, could you identify structural features of Ph.D. education in the U.S. that may impede doctoral students' forays into interdisciplinary research?

The Ph.D. is a research degree, designed to prepare students to become scholars. At the conclusion of the degree program, "the student should have acquired the knowledge and skills expected of a scholar who has made an original contribution to the field and has attained the necessary expertise to continue to do so."[1] Three interconnected features at the core of the contemporary American science doctoral education system have evolved together to push students into specialized disciplinary research. First, academic departments – local manifestations of a discipline – are the primary locus of control for doctoral education. Departments have almost complete control and discretion to set admissions criteria, administer financial support, determine the curriculum, and set the standards for the various requirements that students must complete (exams, proposals, and dissertations). Because departments are designed to foster knowledge within their discipline, and their reputation and resources flow from recognition within the field, it is in the department's interest to foster research that will garner accolades from within the field. This bias towards disciplinary, rather than interdisciplinary, research is expressed in departmental policies. For example, dissertation committee members may be required to come primarily from within the department. Second, students work for and with individual faculty members who can exercise enormous power over students' studies. The most important of these mentors is the advisor, who not only helps students design a course of study but also directs their research. In most of the physical and life science disciplines, students routinely become part of their advisor's research team and build their individual research on work done in the advisor's lab or group. The advisor's earlier research usually provides the intellectual foundation for the dissertation. As a result, a student has strong incentives to follow the research direction set by the advisor, typically located within the mainstream of the discipline. Furthermore, the advisor is often the sole arbiter of whether students have completed sufficient quality work to merit receipt of the degree. Consequently, the advisor can wield enormous control over many aspects of a student's professional life. The recent suicide of Jason Altom, at the time a doctoral student in the Harvard Chemistry department, and the issues he raised in his suicide note, provide dramatic evidence of this point.[2] Last, further ties between advisors and students are created by the mechanisms that provide funding for research. Since World War II, the federal government has deliberately located most federally funded research inside of universities rather than in independent labs, as is the case in many other countries through

---

[1] Council of Graduate Schools, "The Doctor of Philosophy Degree" (Washington DC: Council of Graduate School, 1990), p. 1.

[2] S.S. Hall, "Lethal Chemistry at Harvard," *The New York Times Magazine* (November 29, 1998): 120–8.

research grants awarded to individual faculty members, who in turn use this money to fund their students' tuition and research expenses.[3]

Given these structural features, please discuss the major challenges facing doctoral students who wish to conduct interdisciplinary research

. . .We generated a list of four challenges that are likely to be faced by a student who begins a doctoral program intending to pursue interdisciplinary research. While we list these as separate problems, we see them as interconnected, a point we return to later. Keep in mind that interdisciplinary research can take place between disciplines that are quite close together – chemistry and biochemistry – or disciplines that are dissimilar and distant – sociology and zoology. The challenges of the latter are much greater and as a result each of these obstacles might be more or less problematic for a given student.

Finding an Advisor

The ideal dissertation advisor is supportive, experienced, supplies resources, and socializes the student into the discipline. Choosing an advisor with whom the student can build a supportive professional relationship is perhaps the most critical decision a student makes. In many respects, the student is shaped and changed by the advisor: learning how to identify and think through a problem, how to conduct high-quality research, how to write manuscripts and where to publish them, and so forth. For a student with interdisciplinary interests, a good advisor also needs to understand and share the student's commitment to interdisciplinary research. Identifying such an advisor may not be easy. Interdisciplinary research by students is easiest when the advisor conducts such research him- or herself. However, relatively few faculty conduct interdisciplinary research, although this is impossible to quantify. Some faculty fear the negative consequences of taking up interdisciplinary topics, such as difficulty obtaining tenure, research funding, or peer recognition. This has been identified as a particularly acute problem for untenured faculty.[4] For a student who attempts to conduct investigations outside or beyond the advisor's expertise, additional problems emerge. The advisor may be unable to help the student identify relevant literature and resources. The advisor is likely to be hard-pressed to assist the student in minimizing false starts on research ideas. Relatedly, the student might face the additional hurdle of finding a supportive dissertation committee; faculty from the home department may not support the work, and those in other departments are difficult to identify.

---

[3] R.L. Geiger, "Organized Research Units-Their Role in the Development of University Research," *J. Higher Education* (1990) 61(1): 1–19; P.J. Gumport, "Graduate Education And Organized Research In The United States" in B.R. Clark (ed.) *The Research Foundations Of Graduate Education*, (Berkeley: University of California Press, 1993), 225–60.

[4] P.J. Gumport, "Curricula as Signposts of Cultural Change," *Rev. Higher Education* (1998), 12(1): 49–61; T. A. Heberlein, "Improving Interdisciplinary Research: Integrating the Social and Natural Sciences," *Soc. Nat. Res.* (1998) 1(5): 16.

Mastering Knowledge and Reconciling Conflicting Methodologies

As the amount of knowledge in any field continues to increase dramatically, students must master increasing amounts of information and increasingly specialized techniques – breadth is of necessity sacrificed for depth, producing ever-more specialized researchers. This makes it difficult for students to acquire a sufficiently solid base of knowledge in their own discipline, much less another field, to make significant research contributions. Students with interdisciplinary interests can either attempt to gain knowledge in more than one field or they can collaborate with a researcher outside of their home discipline. Most, of course, do a combination of both, each of which has particular challenges. A student seeking to gain an understanding of multiple disciplines must struggle to master relevant knowledge from each field. It is not enough, however, to recommend that students take basic courses in other fields because they might not even cover topics that cross their interdisciplinary interests in a meaningful way. To illustrate, a chemist working in a neurosciences lab recently told us, "I cannot understand my colleagues' presentations at a level that makes my input useful for the overall goals of the lab. The learning curve is steep, and I simply cannot do all of the reading to get more than cursory knowledge of my labmates' specialties." On the other hand, the student who chooses to collaborate faces the time and emotional strains associated with working with others. Successful collaboration requires power sharing and building trusting interpersonal relationships.[5] Although some maintain that this process is more fulfilling and produces a better end product, experienced researchers cite difficulties in finding partners, coordinating multiple schedules, and negotiating issues such as authorship.[6] Furthermore, incorporating collaborative research into a dissertation is only acceptable in some fields. (In the humanities and many social sciences, solo-authored single works are the standard; a dissertation comprising multiple coauthored papers, as in the sciences, would not be tolerated.) In short, the path to becoming fluent in two or more disciplines is unclear and certainly requires additional time. Environmental scientists would argue that the inherently interdisciplinary nature of ecological research makes collaborative skills especially critical for success. Therefore, graduate programs in these fields need to be especially careful that students obtain experience collaborating. Beyond mastering the concepts and language of another discipline, working in the interstices of two disciplines means conceptualizing and undertaking research in the absence of established and proven frameworks and models. Trying to integrate two disciplines often means resolving conflicts between research paradigms and methods. The research paradigms in different fields (and within some fields) are predicated on different assumptions about

---

[5] W.F. Whyte, "Extradepartmental Enterprise," *Society* (1978) 55(3): 22–5; D.J. Wood and B. Gray, "Toward a Comprehensive Theory of Collaboration," *J. Appl. Behav. Sci.* (1991) 27 (2): 139–62; J.J. Hafernik, D.S. Messerschmitt, S. Vandrick, "Collaborative Research: Why and How?" *Educ. Res.* (December 1997) 31 (5); A. Heberlein, "Improving Interdisciplinary Research: Integrating the Social and Natural Sciences," *Soc. Nat. Res.* (1998) 1 (5).

[6] Heberlein, *op.cit.*, note 4 and Hafernik, *op.cit.* note 5.

what constitutes evidence, what standards of proof are, and what passes for "truth" in the discipline.[7] For example, a well-selected and rigorously constructed case study relying on extensive interview data is seen as a valuable contribution to theory and practice in sociology but is often dismissed in physical science as "anecdotal" and unable to be replicated.[8]

Finding an Intellectual Community

An intellectual community provides valuable socialization and helps contribute to student success.[9] Students need to find faculty to provide intellectual input and fellow students to provide collegiality, emotional support, and a safe arena for formulating and honing new ideas. Working in a non-traditional or emerging field, however, makes it more difficult to develop this type of community. Often the people who would be natural colleagues and collaborators are in several different departments. It is often hard to learn the expertise and interest of scholars who are physically and organizationally distant. This is a particularly challenging obstacle for students to surmount, as there are few mechanisms connecting them to faculty or students in other departments. An interdisciplinary student is vulnerable to feeling intellectually homeless, without a place to share interests and long-term goals.

Overcoming Fears

Graduate students often believe themselves to be significantly dependent on faculty, and perceive their options to be restricted. Consequently, students often act fearfully and avoid taking risks, and, indeed, the stakes are often very high.[10] One fear expressed by students interested in pursuing interdisciplinary research is that their work may have few outlets for publication, and the rewards for such publications are minimal.[11] Colleagues may not value publications in journals outside of the home discipline, which aggravates students' fears about having their work recognized and ultimately becoming employed. Another concern is that the traditional academic job market is now more difficult to enter than in earlier decades, and few signs indicate that this will change. Most positions are

---

[7] Heberlein, *op. cit.* note 4; and B. Laslett, "Interdisciplinary Teaching and Disciplinary Reflexes," *Historical Methods* (1990) 23 (3): 130–2.

[8] M.A. Miller and A.-M., McCartan, "Making the Case for New Interdisciplinary Programs," *Change* (1990) 22(3): 28–35.

[9] L.L. Baird, "The Melancholy of Anatomy: The Personal and Professional Development of Graduate and Professional School Students," in J.C. Smart (editor), *Higher Education: Handbook of Theory and Research* (New York: Agathon Press, 1990) and V. Tinto, "Appendix B: Toward A Theory of Doctoral Persistence," *Leaving College: Rethinking the Causes and Cures of Student Attrition*, 2nd ed. (Chicago: University of Chicago Press, 1993), 230–43.

[10] J. Hahn, "Disciplinary Professionalism: Second View," in R.M. Jones and B.L. Smith (editors), *Against the Current: Reform and Experimentation in Higher Education* (Cambridge: Schenkman Publishing, 1984), 19–33.

[11] Heberlein, *op.cit.*, note 4 and Hafernik, *et.al.*, *op.cit.*, note 5.

located in traditional departments, and students whose work is hard to categorize in traditional ways may be at a disadvantage.[12] Thus choosing an interdisciplinary research topic may be intellectually appealing but not strategic or viable. Of course not all students aspire to careers within academia; many look to industry and government, from which many of the calls for students with interdisciplinary skills emanate. Still, it is difficult to forecast the demand for employees in any field five to ten years in advance – the time line relevant for new graduate students.

Are there potential solutions to these problems?

Tackling [these problems] requires holistic solutions and systems thinking. We believe that small changes, such as encouraging students to take courses in other departments, may not foster truly interdisciplinary research, although they surely are a step in the right direction. If it is to nurture interdisciplinary research, graduate education must be "reshaped," not just tweaked around the edges. One example of a successful effort at nurturing interdisciplinary research is the reorganization of all of the biomedical sciences at Emory University into six interdisciplinary research clusters (such as neurosciences and genetics), instead of more traditional departments. A detailed description of the program and the process of implementation of this change can be found at http://wcer.wisc.edu/gradedforum. While the planning and implementation process was time consuming and required political will and considerable resources, it has resulted in programs that attract high-quality students, have high completion rates, and are highly regarded in the research community. Another example, from ecology, is the Environmental Science and Engineering program at UCLA that has granted doctorates to over 170 students since it was founded in 1973. The program draws faculty from over a dozen departments including sociology, chemical engineering, environmental health sciences, and economics. Interdisciplinary research is fostered both in "problems" courses that provide a collaborative, applied research opportunity and an internship in the field. In both cases, the programs faced significant challenges stemming from the organizational norms of their respective universities.[13]

---

[12] P.J. Gumport, "Feminist Scholarship as a Vocation," *Higher Education* (1990) 20: 231–43 and Miller and McCartan, *op.cit.*, note 8.

[13] R.L. Perrine, "TSCA and the Universities: Educating the Environmental Chemical Professional," *Environ. Professional* (1982) 4(2):187–97.

Thomas Cech, President of the Howard Hughes Medical Institute, Distinguished Professor of Chemistry and Biochemistry, University of Colorado

Gerald Rubin, Vice President and Director, Janelia Farm Research Campus, Howard Hughes Medical Institute

Dr. Cech's work has received the Heineken Prize of the Royal Netherlands Academy of Sciences (1988), the Albert Lasker Basic Medical Research Award (1988), the Nobel Prize in Chemistry (1989), and the National Medal of Science (1995). In 1987, Dr. Cech was elected to the U.S. National Academy of Sciences and also awarded a lifetime professorship by the American Cancer Society. In 2000 Dr. Cech moved to Maryland to be president of the Howard Hughes Medical Institute.

Dr. Rubin is Howard Hughes Investigator and Professor of Genetics and Development. A MacArthur Professor at the University of California, Berkeley, Dr. Rubin is a member of the National Academy of Sciences, the Institute of Medicine, and the American Academy of Arts and Sciences

Drs. Cech and Rubin, you have created a new, interdisciplinary research center, The Janelia Farm Campus of The Howard Hughes Medical Institute. Would you please identify examples that helped you devise your strategies in building this new institutional model for interdisciplinary research?

[...]Two well-known and highly regarded examples are the Medical Research Council Laboratory of Molecular Biology (MRC LMB) and the former AT&T Bell

Laboratories. We have attempted to understand the cultural, organizational, and management features critical to their success. Despite the fact that one of these institutions was a small public-sector biological research laboratory and one a large private-sector electronics enterprise, they share a surprisingly wide range of operating principles.

Individual research groups were small, generally composed of fewer than six individuals. Small group size was considered essential to promote collaboration and communication between groups, as well as good mentoring. Larger projects were often conducted by self-assemblies of smaller groups. Excellent support facilities and infrastructure were provided, enabling individuals and small groups to function effectively and to focus on creative activities. Internal sources provided dependable and generous funding; outside grant applications were not permitted, nor was there any obvious pressure for the work to be of immediate medical relevance or commercial value.

The time made available by the lack of formal teaching, administrative or fundraising activities, as well as the high level of staff support, made it possible for group leaders to carry out experimental work with their own hands. Experienced scientists were also much more approachable and available to discuss scientific ideas and mentor junior scientists. In contrast, it is increasingly rare for a tenured faculty member in a U.S. research university or medical school to spend substantial time working at the bench or even engaging in unscheduled interactions with their colleagues.

At both the MRC LMB and Bell Labs, "success" was defined as solving difficult and important research problems, as opposed to more typical criteria such as publication number, service on editorial boards or speaking invitations. The management of both institutions felt it was their responsibility to be familiar enough with the work of their scientists to be able to evaluate their potential and their accomplishments. They were patient with those who were considered to be very good, but had not yet achieved external recognition. At both institutions, turnover of group leaders was high and tenure was either absent or limited; most group leaders moved on to university positions after five to ten years.

Would you discuss challenges that universities face in institutionalizing integrative research across disciplines, and also possible approaches that can be utilized to overcome these challenges?

[. . .] The universities find it challenging to bring down their internal walls between disciplines. Their biologists are in separate buildings from their physicists; their engineers work in a separate college of engineering; and the nearest medical school may even reside in a different city (University of California Berkeley versus University of California San Francisco; Harvard College in Cambridge versus Harvard Medical School in Boston). Such geographical separation presents a considerable challenge. The spark of transdisciplinary approaches and insights requires "productive collisions" between people in different disciplines, just as atoms and molecules must undergo productive collisions to react. If engineers, biologists, and computer scientists live apart, they need to make an appointment in order to "collide." And individuals are often so highly scheduled that such collisions, even when they occur, are usually of short duration.

Another problem is that the academic culture and promotion system often actively discourage collaboration among research scientists. When scientists collaborate, their unique contributions are often not apparent, a situation that adversely affects their performance review and career advancement. So often we hear of an assistant professor being denied tenure because he or she was a middle author rather than first or last author on too many publications. But order of authorship, necessarily determined in one-dimensional space, does not adequately depict contributions in a multidimensional collaboration. This situation is compounded in the case of inter-disciplinary research where the research advance itself may not be seen as being at the forefront of any of the individual disciplines, the standard-bearers for which generally reside in separate departments that are competing for resources such as new faculty appointments. For example, the development of the field of bioinformatics was greatly delayed because neither biology nor computer science departments saw bioinformatics as an appropriate area of research for their faculty.

In response to these challenges, American universities have constructed buildings for collaborative research separate from the traditional departments. JILA (the Joint Institute for Laboratory Astrophysics) at the University of Colorado at Boulder brings together physicists, chemists and superb instrument engineers and fabricators, and has garnered two recent Nobel Prizes. BIO-X at Stanford University has equal doses of biology, physics, and engineering. Princeton University's Center for Integrative Genomics merges molecular biology, chemistry and physics. The Skaggs Institute of the Scripps Research Institute is especially strong at the interface of chemistry and biology, whereas the new Broad Institute will bring Massachusetts Institute of Technology, Whitehead Institute and Harvard scientists together to apply genomics and chemistry to problems of medicine. Of course, when you bring energetic, creative people together, it is not surprising that great things happen. But the measure of success for these interdisciplinary institutes needs to be "Is the whole greater than the sum of the parts? And, if so, by how much?"

The challenges of promoting innovative interdisciplinary biological research are routinely solved – not in academia, but in biotechnology companies. There, if product development requires a team of bioinformaticists, chemists, cell biologists, pharmacologists, and chemical engineers, the company wastes no time in bringing them together. Because they live under a single roof, the challenge of dealing with geographic separation is mitigated. And the team benefits from being faced with a well-defined target for their work. Perhaps academic institutions, which currently select and train their scientists to be fiercely independent, could benefit from adopting a bit of this attitude in their interdisciplinary science.

Would you discuss the paradigms for integrative science research that HHMI's Janelia Farm employs?

The Janelia Farm Research Campus ... will aim to identify important biomedical problems for which future progress requires technological innovation and then foster the self-assembly of integrated teams of biologists and tool builders who seek to break through the existing barriers. This in turn will require a much stronger focus on developing new tools – experimental methods, computer software, and scientific instruments – than is typical in most academic research centers. The scientific problems we choose to pursue will drive the choices of tools we seek to develop; the software and instrument development activities will work in close concert with, and support, the ongoing experimental work. The group leaders will be recruited through an open international competition. Individual research groups will be capped at six people, mostly postdoctoral fellows and technicians, but extensive shared core facilities will be provided; group leaders will be able, even expected, to continue to work at the bench. A single laboratory building will allow about 400 individuals from different disciplines to work together without departmental walls.

We believe that having HHMI provide all, or nearly all, funding for activities at Janelia Farm is essential for creating the interactive, interdisciplinary and collaborative research environment we envision. Only in this way can we provide scientists with the freedom from distractions that will allow them to participate directly in experimental work and the freedom to tackle important problems that might have too low a probability of success – or not yet be well-enough defined – for a typical grant application.

An assistant professor's freedom to select a research problem is generally limited to problems that can garner external research support. Not only are grant-giving agencies notoriously risk-averse, but the work will probably be carried out by postdoctoral fellows and graduate students who are under pressure to obtain publishable results within a short time span. Thus much emphasis is placed on quantity and certainty of output at the expense of originality and potential impact. Similarly, research in biotechnology companies must support the business plan. Although these funding models are appropriate for the vast majority of biomedical research, they have two major limitations. First, proposals for more adventuresome projects, even those that may have enormous impact if successful, have traditionally fared poorly. This is especially true for non-hypothesis-driven research aimed at developing new research tools. Second, the ability to move quickly to take advantage of unforeseen targets of opportunity is severely constrained.

With small individual groups, Janelia scientists will have to collaborate to tackle a big project. The group leaders will be valued and evaluated on the basis of their contribution to collaborations, supportive interactions, and mentoring of other scientists as well as their individual achievement, overcoming the most pervasive obstacles to collaboration in academia. Decisions about renewal of appointments every five to six years will include an evaluation by an external panel of experts who will interact with the scientist being reviewed, rather than simply judging from publications and written reports. The process and the expected level of performance will be similar to those now applied to the review of HHMI investigator appointments. The external opinion will be weighed along with an internal evaluation of the contributions to collaborations and mentoring. Although group leader appointments will be untenured, there will be no limit on the number of terms for which a scientist can be reappointed.

It is critical that Janelia Farm be able to evolve as science evolves. Thus, some group leaders will be asked to leave not because the quality of their work has slipped, but because their expertise no longer fits with the mission of the campus. Such group leaders will be allowed to transfer their HHMI appointment to any of our host institutions that wishes to receive them, and after another five years they will be evaluated on an equal footing with the other investigators. This "soft landing" should make having an untenured position quite palatable.

Academic institutions have one key resource that biotech companies do not share: an abundant supply of graduate and undergraduate students as well as postdoctoral fellows. Interdisciplinary research benefits greatly from students and postdocs, who are not so mired in disciplinary dogma and transgress boundaries rather effortlessly. From the other side, students benefit greatly from interdisciplinary research; they find it intellectually invigorating, they enjoy the social dynamics of a team approach to problems, and they derive excellent preparation for a possible future in a more interdisciplinary academic environment or in the biotech industry. Thus, although Janelia Farm will not be a degree-granting academic institution, it will provide opportunities for students from partnering universities throughout the world to carry out all, or part, of their thesis research on site. Furthermore, a robust visiting scientist program will allow visitors with appropriate interests and expertise to participate in ongoing projects with Janelia's resident scientists and also serve to make the tools created at the farm broadly accessible and promote their dissemination.

Another area where Janelia Farm aims to have an impact is increasing the participation rate of women in the highest-level biological research. Unlike the case of under-represented minorities, the pipeline of talented young female biologists is full, yet women still make up less than 15% of full professors in departments of biological sciences at research universities. It is widely agreed that a major factor in the declining participation of women is the conflict between an academic career in science and the demands of assuming primary childcare responsibilities. Janelia Farm is well positioned to provide a supportive environment for women scientists. Most importantly, the lack of nearly all professional obligations not directly related to research, and the provision of high-quality on-site infant and child care, should allow the time needed for both high-level research and family life.

The majority of scientists at Janelia will be focused around two or three broad goals or missions. We call these "1,000 person-year projects," and ask what biological goal would be worth the effort of 100 scientists working for 10 years with abundant resources, with instrumentation and computational tools being created in-house to further the project. A series of workshops held earlier this year, each with about 30 participants, has helped evaluate potential initial research objectives (see http://www.hhmi.org/janelia for a current description of these workshops). We are particularly interested in identifying research areas that may be underpopulated not because they are seen as unimportant, but because they are considered too challenging for traditional funding mechanisms and career structures or because they require an interdisciplinary approach not feasible in a university setting. It is important to emphasize that the initial objectives are meant to provide some focus but not to constrain the creativity of individual scientists to follow opportunities as they arise.

Clearly, interdisciplinary research will take many forms over the coming decades as more scientists and institutions become committed to this approach. A key related issue is whether undergraduate and graduate school curricula will be able to change to support a more integrated approach to biological research problems.

Chapter 10
What Are the Architectural Implications of Integration?

Robert Venturi, Pritzker Prize winner in Architecture and founding principal of the firm Venturi, Scott Brown and Associates

Robert Venturi studied at Princeton, then worked for Louis Kahn before establishing the Philadelphia firm with John Keiser Rauch (1930–) that became Venturi, Rauch, Scott Brown and Associates (1958). His buildings include the Sainsbury Wing of the National Gallery, London (1991) and the University of Michigan Life Sciences Institute (2004). He received the Pritzker Prize in 1991.

M.E. Brint et al. (eds.), *Integrated Science*, DOI 10.1007/978-0-387-84853-2_10,
© Springer Science+Business Media, LLC 2009

How do you approach the design of a scientific research laboratory?

I should first warn you that I come to this subject as a practicing architect. My generalizations are really pragmatic responses to everyday experiences. The most important thing, in my opinion, about the architecture of the Research Laboratory is that it is generic.

What do you mean by "generic"?

[The] kind of generic architecture [I have in mind] concentrates on three... characteristics:

– The element of FLEXIBILITY – Spatial and mechanical – that is promoted inside,
– The imageries of SETTING & PLACE that are accommodated inside,
– The elements of SYMBOLISM &ORNAMENT – permanent and/or changing – that enhance imagery on the outside.

As a form of architecture, how does flexibility work in terms of the functional use of the research laboratory today?

...No longer does form follow function. But that's not
ambiguous enough

> **FLEXIBILITY-FOR TODAY**

it is rather that *form accommodates functions*: functions that are inherently changing, as they are complex and contradictory.

So form should not express the function of the building?

Functions [should be] accommodated rather than expressed. This is more and more relevant in science buildings in particular and in architecture in general, to accommodate change that is more characteristically revolutionary than evolutionary and that is dynamically wide in its range: spatial, programmatic, perceptual, technical, iconographic. In our time, functional ambiguity rather than functional clarity can accommodate the potential for "things not dreamt of in your philosophy."

In addition to the requirement of a flexible form that can accommodate changing functions, you mentioned place and setting as a second element of generic architecture. Where would you find an example of the importance of setting and place in architectural design?

Where the scale of the architecture is physically generous, to create an aura of generosity as well as accommodate the dynamics of flexibility.

How do you distinguish between setting and place? Do they serve different general functions?

– For concentration: *Setting*.
– For communication: *Place*.

Could you give us a few specific examples?

> *Setting – as* background for work and focus – alone or with a group of colleagues.
> *Place – as* opportunity for meeting – meeting that is incidental rather than explicit. (Academics are perverse: If your architecture explicitly pronounces a place for interaction, they might not use it; a certain ambiguity concerning the function and nature of the space is perhaps essential here.)
> *Setting – as* more or less local.
> *Place – as* signification of a whole (or perhaps a suggestion of a whole): a Community, an Academic Community within a campus – so you function alone *and* in a community.
> *Setting – a* probably messy space: for the clutter of creative action, analytical, intuitive, physical.
> *Place – a* mostly orderly space: for re-creation, so to speak.
> *Setting – within* the consistent structural order – the generic order – of the loft, accommodating variety within order: spatial, perceptual, functional, mechanical.
> *Place – as* an exception to the rule of the consistent order of structural bays, accommodating a special space.
> *Setting – accommodating* change and dynamics in the lab: Wow, look at this!
> *Place – accommodating* a permanent ambience where one anticipates the comfort of the familiar, but where there can be the surprise of the unfamiliar: "Wow! John, of all people, said something brilliantly relevant this morning as we chatted!"
> *Setting – neutral*, recessive architecturally, to diminish distraction. Artists' studios are in lofts not essentially because artists are poor, but because they feel they can't create a masterpiece in someone else's masterpiece. A setting for inspiration *and* perspiration: artists and scientists are not priests performing rituals.
> *Place – imageful* architecturally, to create amenity and identity.

There seems to be a tension between setting and place in this generic architecture?

A technical IRONY regarding the significance of Place . . .: [An] imageful place that is local might be more essential than ever in our era of electronic communication to and from all over, in our era of networking.

An aesthetic IRONY is that accommodation to Place in this architectural context promotes a rhythmic exception within generic order and thereby creates aesthetic tension.

The final element of generic architectural design is symbolism and ornamentation. What role(s) do these aspects play in generic design?

> *Symbolism is* the attendant flourish within Generic Architecture that eloquently breaks the consistent order – that consistent rhythmic order – on the outside (and some places inside), thereby creating aesthetic tension.
>
> *Symbolism – involving* ornament, sign, iconography – on the outside. Traditionally evident, as in the cupola of Nassau Hall at Princeton, the *portone* with *stemma* of the Italian palazzo, the sign atop the mill, the hieroglyphics all over the Egyptian temple.

> **SYMBOLISM &
> ORNAMENT-THAT ENGAGES
> ICONOGRAPHY ON THE
> EXTERIOR**

The difference between outside and inside – at least in the lab building, and especially for the academic lab building – is very relevant where architectural consistency and neutrality of the workplace inside is counterbalanced by explicit symbolic content on the outside that acknowledges the significance of the institution as a whole.

Another role of exterior *Symbolism* and *Ornament* in the generic Academic Lab Building is that which acknowledges, accommodates, and enhances the context of the architectural campus.

Your emphasis on iconography and ornamentation sounds like a reaction to or departure from modernist approaches?

Modernism employed industrial engineering imagery as an architectural aesthetic, sometimes called the machine aesthetic, along with a Minimalist-Cubist abstract aesthetic. It did this via an adaptation of the vernacular vocabulary of the industrial loft – essentially the American generic loft of the turn of the century and this exemplified a wonderful/valid historical-architectural evolution (or revolution) formally acknowledged in this country when Dean Hudnut invited Walter Gropius to Harvard.

The neo-Modernist movement in the architecture of today involves a revival of engineering expressionism more explicitly ornamental than before. It involves, in the end, an ironical architectural vocabulary based on industrial imagery as industrial *rocaille*, an imagery that is now around 100 years old and in our admittedly post-industrial age no more current or relevant than – and no less historical than – that imagery of the Classical orders of the Renaissance that are 500 years old. Everyone agrees that the Industrial Revolution is dead, but few architects acknowledge that Electronic Technology is what can be fundamentally relevant for architecture today. Electronic Technology, combined with generic order, can enhance, can indeed signify, an iconographic dimension – an iconographic dimension that is for *now* unlike the spatial-structural dimension that was for *then – but* an iconographic dimension nevertheless with a vivid tradition behind it.

So industrial and engineering/structural imagery of space is incidental for now, while an ornamental/symbolic imagery of appliqué is valid for now.

Does your firm employ these elements in designing scientific laboratories?

The scientific laboratory buildings designed by our office...illustrate variously qualities of the generic loft, whose interior flexibility accommodates programmatic, spatial, and mechanical evolution over time and whose exterior ornamentation, within the consistent rhythmic composition of the loft, accommodates symbolic dimensions appropriate for a communal academic building. Exceptions to these forms of order deriving from incidental interactive spaces enrich the composition of the whole inside and out.

The designs of buildings like the Lewis Thomas Laboratory for Molecular Biology at Princeton University [above], represent work we have done in association with Payette Associates Inc. of Boston who are most significantly responsible for the major interior spatial-mechanical-programmatic elements of the architecture. And it is Jim Collins of that firm with whom we have had the pleasure and honor of working in the last decade.

Could you tell us about how the Lewis Thomas Laboratory project unfolded?

... [Our] firm had successfully completed a critically acclaimed building on campus, Gordon Wu Hall, and had recently developed the campus design guidelines for the area of campus in which the proposed building was to be located. They were, therefore, an obvious choice to develop the new building's exterior.

At President Bowen's urging, ... [we] were invited to submit a proposal and were chosen to lead the project. Payette, because of its expertise in technical buildings, would develop the program and design the interior spaces. VRSB, with a budget of $2.5 million, would work with the facade and site plan, integrating the new building with its traditional, neo-Gothic surroundings. Each organization had a clearly bounded sphere of influence. As Thomas Payette put it, any design on the outside was VRSB's ultimate decision, and design on the inside was Payette's ultimate decision. Payette Associates would be the architects of record and would have overall responsibility for the project management and documentation. By the very nature of this collaborative endeavor, teamwork was stressed over individual inspiration.

It was determined that the building would be developed around the "generic" laboratory vision of the new chairman of the Department, Arnold Levine. . . . As the majority of the building's occupants were yet to be recruited, the program and the concept of the building organization had to be developed on a generic basis by a small team. Arnold Levine, his associate Tom Shenk, and I, as Payette project architect, formed the core of the programming effort. [We decided on an] "open lab" concept.

This open lab concept was not new. The large open lab had been successfully used as early as 1965 in Louis Kahn's design of the Salk Laboratories in La Jolla, California, but Salk and a few other notable laboratories were the exception. At the time the Lewis Thomas Lab was being planned, the dominant approach was to design discrete, small laboratories reflecting the hierarchical nature of "senior scientists" and "junior assistants." Most scientific laboratories did little to encourage interaction among scientists, either.

How do you incorporate the changing landscape of science when you design buildings like the Lewis Thomas Laboratory?

Because the Lewis Thomas Laboratory buildings involve research in a rapidly progressing field, the goal of planning for unpredictable change continually challenged our design approach. Indeed, in any laboratory design, planning for the future is nearly as important as is planning for the present. Since molecular biology is evolving almost daily, there is constant pressure to adjust to ever changing standards and trends. Understandably, because a great deal of flexibility was expected in order to meet these unforeseen future challenges, these attributes were fundamental to the project's viability.

Throughout the entire design process, allowances needed to be made, predictions ventured, and safety issues assiduously addressed. Whereas some elements of the program appeared to have been fixed through program space requirements, in actuality, the planning was structurally tight, while allowing for a great deal of flexibility. Space above each lab was planned to allow the mechanical and electrical systems to be moved easily and installed elsewhere without disrupting the lab below. Large, open lab spaces could be reconfigured and subdivided in the event that existing laboratory needs changed. Because much of the electrical and utility space was run up through shafts at either end of the loft-like laboratories, rather than up through each individual laboratory station, a great deal of flexibility was incorporated.

If you create an interdisciplinary science space, will it encourage interdisciplinary behavior among and between scientists?

Admittedly, although the creation of a viable and effective laboratory involves numerous technological constraints, at Payette Associates, we know that research is ultimately about people. People like choice, thus we sought to provide a variety of different kinds of spaces within the building: closed, quiet spaces for contemplation and individual work; open public spaces for spontaneous activity and discussion outside the laboratories; and research space that also encourages the continuous exchange of information between investigators.

A generous staircase, for instance, generally invites exchanges between and among floors, as people constantly pass each other and relate their laboratory experiences (See figure). Traffic patterns can be tightly controlled when a single corridor functions as a main thoroughfare. At the Lewis Thomas Laboratory, offices were grouped in clusters rather than in separate laboratories to reinforce the strong sense of community felt among the members of this interdisciplinary group. Blackboards were strategically placed to invite spontaneous interactions and impromptu gatherings. In essence, by manipulating the frequency with which researchers exchange information, architects can effectively promote the sharing of knowledge through the sharing of space, resources, and facilities.

Do you think that architecture changes human behavior so that you can influence the way science is done?

On this level we architects and these scientists are very much the same. We are interested in discovering, understanding, and influencing life.

In recent years, architectural theorists have disagreed over whether or not the social behavior of a building's users is influenced, even determined, by the physical environment in which that behavior occurs. Proponents of this influence – architectural determinists – believe that designers can direct social behavior through their work. Using the Lewis Thomas Laboratory at Princeton University as a case study, we can positively attest that there is a direct correlation between the work environment and the workers' intellectual and physical activity. Furthermore, a sense of order, continuity, and cohesive structure are all expected to have a positive impact on the way scientists relate to their surroundings.

We can, as architects, through our definition and manipulation of space, create a positive and nurturing environment for research scientists. Spatial planning can foster the paradoxical factors inherent to the research laboratory: innovation and replication, discussion and reflection, teamwork, and competition. . . .Many in the field continue to view research labs as highly controlled environments supported by intense, space-consuming mechanical systems. They cite examples where major science has been accomplished in the most inhospitable of places. Given our own experience in building for the scientific community, we are inclined to believe that the very opposite is true. Perhaps Dr. Jonas Salk, who discovered the polio vaccine and founded the research institute that bears his name, summed up our theory most accurately when he discussed Louis Kahn's design of his institute:

> My ambition was to optimize the functioning of the human mind, to deal with the issues and questions with which the human mind is concerned. I wanted to create something that would influence the realm of the mind - the minds of those who would gather here to carry on this kind of work. I was seeking a retreat atmosphere for reflection and work, away from the business and noise of the world. . .Architecture is used here. Some people pursue science for human use, in contrast to science for the sake of science. This architecture is for human use, to serve a purpose.[1]

---

[1] Salk is quoted in Michael J. Crosbie, "The Salk Institute," *Progressive Architecture* (October 1993): 47.

Claire M. Fraser, Director, Institute for Genome Sciences (TIGR)

Dr. Fraser has published more than 160 articles in scientific journals and books. Before becoming TIGR's President in 1998, Dr. Fraser was the Institute's Vice-President of Research and Director of its Microbial Genomics Department. She has received numerous academic and professional honors, including professorships in both microbiology and in pharmacology at The George Washington University

In your work helping to design the new building for the Institute for Genomic Research in Rockville, what considerations regarding interdisciplinary science did you have in mind?

[In terms of physical space for interdisciplinary research,] we have ended up where we are by accident, but at least we were smart enough to see a successful formula. We brought people together with very, very different backgrounds, people who had formal training in molecular biology, microbiology, computer science, mathematics and essentially put them all together in the same space. And they were energized by the possibilities of what could happen if they were able to work together. Our initial lab was a large, open lab which created an opportunity every single day for these people, who might not see each other if they were at a university setting to come together, to learn each other's language, to brainstorm about how to solve some problems of interest collectively. This really is the formula for success. One cannot over-emphasize the importance of physical proximity, of bringing people together, if you truly want to create an interdisciplinary environment.

References

A. Gore, "Address," Presented February 12, 1996, in Baltimore, MD, at the AAAS Annual Meeting, *American Association for the Advancement of Science* printed Apr 12, 1996. Copyright by AAAS. Original address can be found at: http://guilde.jeunes-rhercheurs.org/Reflexions/Pre1997/Archives/AlGore.html. With permission.

B. Alberts (Former President (1993–2005), U.S. National Academy of Sciences). Most of the remarks by Dr. Alberts are from TWAS Newsletter, Vol. 14 No. 4, Oct–Dec 2002 http://www.ictp.trieste.it/~twas/pdf/NL14_4_PDF/07-Alberts_low.pdf. Other remarks are from "Transcription," *Rensselaer Presidential Colloquy*, 9/10/2004, RPI Center for Biotechnology and Interdisciplinary Studies. With permission.

E. Zerhouni, Most of the remarks by Dr. Zerhouni are from "Policy Forum" *Science* (3 October 2003) 302 at www.sciencemag.org; Other remarks are found in "NIH Launches Interdisciplinary Research Consortia," *NIH News*, September 6, 2007, http://www.nih.gov/news/pr/sep2007/od-06.htm; Transcription, Rensselaer Presidential Colloquy, 9/10/2004, RPI Center for Biotechnology and Interdisciplinary Studies; and 'Tangible Benefits Not Created in a Vacuum," *The NIH Record*, (February 3, 2004) 56(3) at http://nihrecord.od.nih.gov/newsletters/02_03_2004/ story01.htm. With permission.

J. J. Duderstadt, "*Commercialization of the Academy: Seeking a Balance between the Marketplace and Public Interest*," in Buying In or Selling Out: The Commercialization of the American Research University, Edited by Donald G. Stern, Piscataway, NJ, Rutgers University press, 2004, pp. 56–74. With permission.

S. Aronowitz, "*The Corporate University and the Politics of Education*," The Educational Forum, (Summer 2000) 64 (4): 332–9. ©Kappa Delta Pi, International Honor Society in Education, 2000. With permission.

D. Kirp, "*The New U*," The Nation, April 17, 2000 posted March 30, 2000 at http://www.thenation.com/doc/20000417/kirp. With permission.

H. Riggs, "*Not So Different After All: Academic and Industrial Leadership in the 1990s*" Occasional Paper No. 29, Association of Governing Boards of Universities and Colleges. With permission.

W. Haseltine, "Transcription," *Rensselaer Presidential Colloquy*, 9/10/2004, RPI Center for Biotechnology and Interdisciplinary Studies. With permission.

S. Brint, "Creating the Future: 'New Directions' in American Research Universities" *Minerva* (2005) 43: 23–50, with kind permission from Springer Science and Business Media.

P. Grobstein "A Vision of Science (and Science Education) in the 21st Century: Everybody 'Getting It Less Wrong' Together." Serendip. 15 March 2003. http://serendip.brynmawr.edu/sci_cult/imsa/imsatalk.html. With permission.

W. Wulf, "The Urgency of Engineering Education Reform," excerpted from *The Bridge*, Vol. 28, No. 1, 2002. Reprinted with permission of the National Academy of Engineering.

D. Kennedy, "Science and the Liberal Arts College," *CUR Quarterly*, (Sept., 2001): 16–20. With permission.

R. Colwell, "Rethinking the Rules to Promote Diversity," *Presidential Symposium on Diversity*, Boston, MA, Aug 18, 2002. With permission.

S. Tobias, "The Professional Science (Math) Master's Degrees: History and Prospects," *The Communicator*, Council of Graduate Schools, Washington, D.C. 39 (6), (July 2006): 3–6. © Council of Graduate Schools, Washington, D.C. With permission.

D. Baltimore, "Promoting Quality and Creativity in Faculty and Students," AAMC talk, March 2004 – Council of Academic Societies. © David Baltimore, 2004. With permission.

C. M. Golde and H. A. Gallagher, "The Challenges of Conducting Interdisciplinary Research in Traditional Doctoral Programs," *Ecosystems* 2 (1999): 281–285, with kind permission of Springer Science and Business Media.

T. Cech and G. Rubin, "Nurturing Interdisciplinary Research," *Nature: Structural and Molecular Biology*, 11(12), (December, 2004). With permission.

R. Venturi, "Thoughts on the Architecture of the Scientific Workplace: Community, Change and Continuity," in *The Architecture of Science*. edited by Peter Galison and Emily Thompson, Cambridge, MA: The MIT Press, pp. 385–398, ©1999 Massachusetts Institute of Technology, by permission of the MIT Press.

C. Fraser, Transcription, *Rensselaer Presidential Colloquy*, 9/10/2004, RPI Center for Biotechnology and Interdisciplinary Studies. With permission.

Index

Printed in the United States